The Economic Dynamics of Environmental Law

The Economic Dynamics of Environmental Law

David M. Driesen

The MIT Press
Cambridge, Massachusetts
London, England

This book was set in Sabon by The MIT Press.
Printed on recycled paper and bound in the United States of America.

Library of Congress Cataloging-in-Publication Data

Driesen, David M.
  The economic dynamics of environmental law / David M. Driesen.
  p. cm.
  Includes bibliographical references and index.
  ISBN 0-262-04211-8 (hc. : alk. paper) — ISBN 0-262-54139-4 (pbk. : alk. paper)
  1. Environmental policy—United States. 2. Environmental law—United States.
  I. Title.
HC110.E5 D74   2003
333.7'0973—dc21   2002071764

Contents

Preface   vii

1 Introduction   1

I Static Efficiency   11

2 Cost-Benefit Analysis and Allocative Efficiency   15
3 Free Trade and Environmental Protection   33
4 Cost Effectiveness and Instrument Choice   49

II The Economic Dynamics of Innovation   73

5 Innovation's Value   75
6 Economic Incentives for Innovation   93
7 Decision-Making Structures   107
8 The Shape of Environmental Problems: Growth, Decentralization, and Change over Time   123

III Toward Economic Dynamic Reform   137

9 Privatizing Environmental Law   139
10 Making Public Environmental Decisions More Fair and Effective   163
11 Improving Regulatory Design to Stimulate Change   183

IV Toward Economic Dynamic Legal Theory   203

12 Static Efficiency Reconsidered   205
13 Conclusion   215

Notes   217
Index   255

Preface

This book involves an effort to address the heart of how we think about law and policy, especially environmental law and policy. As a lawyer with the Natural Resources Defense Council (NDRC), I was struck by how often academic concepts of efficiency strongly influence government decisions about environmental regulation, even under statutes that are based on other concepts, like the protection of public health. When I joined the faculty of Syracuse University College of Law in 1995, I began to study the academic thought on efficiency in more detail, producing a number of articles evaluating various aspects of efficiency-based concepts, especially as applied to environmental law. The research conducted to prepare this work showed me that efficiency concepts dominated the thinking of many academics, including the thinking of some who (relying on a rich literature questioning efficiency concepts) explicitly took exception to some of the tenets of Chicago School law and economics.

Many academics and practitioners expressing dissatisfaction with Chicago School Law and Economics seemed to yearn for an alternative approach, sensing that efficiency-based thinking leaves something important out. This book seeks to provide an alternative by developing a concept of economic dynamics and showing in detail how it might reshape thinking about environmental law and policy. The last chapter explains, with some examples from outside the realm of environmental law, how the economic dynamic theory might contribute to law and economics more generally.

Any book focusing upon such central concepts faces a daunting set of expectations, because of the multidisciplinary nature of environmental policy, and of law and economics. Economists, academic lawyers, ecologists, and political scientists all expect to see the rich work in their disciplines that might be relevant explicitly addressed. But public policy makers will hope for a nice, straightforward exposition, rather than a long, complicated review of others' concepts and work. I have tried to meet these conflicting expectations as best I can by citing the work most relevant to this book's concerns and leaving out reference to works that relate to the topic peripherally, while keeping the exposition as straightforward as I can.

Syracuse University College of Law under Dean Daan Braveman supported this book with several summer grants and a one semester research sabbatical. Alissa DiRubbo, Wendy Scott, and Rosemarie Romano, and the rest of the staff of the H. Douglas Barclay Law Library at Syracuse University always responded promptly and helpfully to my never-ending requests for new materials and reference services. Chris Ramsdale, Theresa Coulter, and Peter Rolph provided very useful assistance in proofreading and preparing the index.

I am very grateful to a number of people who read drafts of this book and made useful comments. These include Alan Mazur, John Dernbach, Stephen Meyer, David Popp, and a number of anonymous reviewers. Many of those who attended presentations of this book during its early stages at the Environmental Law Institute, the Association of American Law Schools (AALS) annual convention, and at the meeting of the AALS's law and socio-economics section also provided helpful suggestions. Nicholas Ashford, in addition to offering valuable suggestions during two of these presentations, provided much appreciated advice and counsel. I have also benefitted from my colleague Robin Malloy's thinking on law and economics and his advice on a variety of matters.

I write this book out of a conviction that we need to ask different questions about environmental policy and about law and economics than we have been asking. I doubt that I will convince all my readers that I have all of the answers to the questions I raise, but I do hope that this book helps refocus the ongoing debates about both the future of law and economics and the reform of environmental policy in productive ways.

The Economic Dynamics of Environmental Law

1

Introduction

The idea that government should serve a public interest largely distinct from private economic interests has suffered a decline in prestige, at least among policy-makers and academics, in the face of an assault from the law and economics movement.[1] As a result of law and economics' growing influence, analysis of public law increasingly relies upon the tools of neoclassical economics, and recommendations for reform often follow the prescriptions of academic economists.[2]

This economics-based regulatory reform agenda has become quite salient in policy-making circles, because it generally supports the views of regulated industries, which have successfully championed economic reform of regulation. In the environmental area this agenda includes increased use of cost-benefit analysis (CBA) to determine the goals of environmental regulation and "economic incentive" measures, especially emissions trading, as the means of meeting environmental goals. Internationally, the law of international trade has become a means of reforming domestic laws protecting the environment and natural resources. A climate of ideas hospitable to economic prescriptions for reform has convinced many policy analysts to support all or part of the reform agenda. This climate of ideas helps shape the training of future lawyers and policy-makers, who will influence how society crafts and implements future laws.

Many of those opposed to the regulatory reform movement defend traditional regulation. Much of this defense contrasts ideal efficiency, the major goal of neoclassical microeconomics, with the real world effects likely to flow from specific economic reforms.[3] It also includes refutation of negative claims about traditional environmental law. Opponents of regulatory reform usually argue that the economic reforms advocated will, as a practical matter, impede realization of public interest goals. Only a few have either defended the idea of the public interest explicitly or raised serious questions about whether efficiency is an important or appropriate goal. And fewer still have sought to provide an alternative direction for regulatory reform. In short, many opponents of pro-business regulatory reform have defended the status quo rather than advanced new ideas.

The dearth of successful analytical alternatives to neoclassical analysis matters a lot, given the dominance of law and economics thinking. The regulatory reform agenda in environmental law forms a small part of a larger market emulation project stimulated, in part, by the law and economics movement. Increasingly, government reform aims to emulate, rather than check, free markets.[4] Government may privatize prisons, public schools, and waste disposal, thus creating a free market where none existed.[5] Government has at least partially deregulated industries that it has traditionally regulated closely, such as public utilities and airlines. And the federal government has largely abolished welfare, requiring the poor to participate in the free market as employees, and considered turning to "faith-based" private organizations to deliver important social services. Where the government remains in a traditional role, it often seeks to "reinvent" itself as an imitator of free market virtues. Thus, the Internal Revenue Service, for example, seeks to emphasize public service in the mode of a large corporation, rather than enforcement of tax law. In all of these cases (and many more), the government tries to have a public service performed in a way that emulates some free market virtue.

Such widespread pursuit of market emulation projects suggests the need for careful thinking about precisely which free market virtues government should emulate. But this kind of analysis actually is in short supply. This book begins to remedy this situation, supplying some detailed thinking about the implications of emulating different free market virtues in one area, that of environmental law. By so doing, this book introduces a broader critique of law and economics that can serve as a foundation for similar analysis in other areas.

Law and economics has been based on two fundamental insights. First, because economic incentives are powerful, thinking about them may help explain how law and policy actually work. Most legal scholars have assimilated this insight almost unconsciously. Nearly everyone uses it, but relatively few scholars think about it much. Second, one can always critique any law in terms of economic efficiency.

In the environmental area, efficiency-based analysis has dominated. Typically, analysts ask whether environmental regulation makes efficient use of private sector resources used to comply with environmental laws. They often find environmental law inefficient and then recommend a reform to enhance environmental law's efficiency. Indeed, scholars sometimes do this unconsciously, for example, by inferring the existence of poor priority setting from evidence of allocative inefficiency.[6]

This choice of efficiency-based analysis seems natural to economists. Economists commonly use an efficiency concept to model a perfect market.

Strangely, however, policy analysts and academic lawyers have not asked whether the efficiency model reflects the most imitation-worthy features of free markets for public policy purposes. Since economists do not claim that the free market actually conforms to the model of a perfect market, the law and policy analysts' selection of the ideal efficiency model seems rather odd.[7] The choice of perfect efficiency allows an analytical concept, rather than a real description of actual free markets, to drive the market emulation project. Analysis compares the efficiency of regulatory systems to the efficiency of ideal markets, rather than to the performance of real markets. Of course, no system, not even the free market, matches the efficiency of this ideal.

More fundamentally, the normative value of efficiency is controversial, even among economists. Many academics employing a static efficiency framework make rather modest claims for the value of efficiency analysis. They point out that efficiency is always a relevant consideration and therefore a potentially fruitful method of analysis in a wide variety of areas.

And many thoughtful academics have doubts about whether efficiency-based analysis provides a sufficient basis for all policy-making. Indeed, when the 104th Congress was considering converting the entire regulatory system to standards based on CBA, many prominent economists signed a statement stating that CBA is "neither necessary nor sufficient for designing sensible public policy."[8] Efficiency-based analysis may not provide a sufficient basis for either critiquing existing environmental protection or coming up with sensible reforms.

Legal academics have been at the forefront of making bolder claims for efficiency-based thinking. Richard Posner, a University of Chicago Law

School professor and a federal judge, has argued that pursuit of economic efficiency increases wealth. He has gone further and argued for the primacy of wealth enhancement as a value undergirding the legal system.[9]

Economic dynamic analysis calls into question the notion that static efficiency is important enough to merit the kind of obsessive attention it has received, even under a wealth maximization theory of law. After all, technical progress, a response to economic dynamics, has provided a large contribution to the growth of industrial economies relative to that provided by increased inputs of capital and labor.[10] This means that frequently inefficient creativity and experimentation play major roles in creating wealth.[11]

Indeed, even that part of economic growth that comes from increased inputs of capital and labor responds in part to economic dynamics. And analysis of the desirable amounts of these inputs in economic theory involves questions of macroeconomic intertemporal efficiency, rather than only static microeconomic allocative efficiency.

Economists define allocative efficiency in terms of matching supply and demand for a given technological state, an aspect of microeconomic theory. I refer to efficiency concepts that assume an unchanging technological state as static efficiency concepts. Economists generally address determinants of rates of innovation and economic growth as a macroeconomic topic separate from static efficiency concerns. Indeed, the proposition that perfect static allocative efficiency contributes more to economic growth than a less efficient allocation is controversial in the economics profession. A debate has raged between economists about whether perfect competition is good for economic growth, even though all concede that perfect competition is necessary for optimum static allocation of goods and services.[12] Perfect competition (a pre-condition for static efficiency) benefits consumers by lowering prices to the point that profits diminish. But lower profit levels may make investment in technological innovation difficult, thus retarding economic growth.[13]

Equilibriums come and go as the economy grows and changes over time.[14] In the environmental area, a cost-benefit based regulation alters the equilibrium an agency seeks to capture in a CBA-based rule as soon as the regulation becomes effective. The regulation sets off a competition to provide the environmental goods and services needed for compliance at the lowest possible cost. This competition establishes a new equilibrium dif-

ferent from that envisioned by the agency during the rulemaking proceeding producing the regulation. Temporary static equilibriums may simply
not matter very much in the long run.

If the economy changes drastically and grows in scale, then the relationship of this growth to the earth's carrying capacity, the issue of scale,
becomes more important than efficiency, even if one accepts a wealth creation goal. Efficiency, after all, simply allocates fixed resources. It neither
augments resources, as economic development does, nor diminishes available resources, as natural resource depletion does.

This does not mean that efficiency plays no role in wealth creation or has
no value. But static efficiency may pale in significance when compared to
economic dynamics.

The kind of static efficiency that economists use to model markets has
little relation to very important free market virtues one might wish to emulate. Surely, many people admire the free market's tendency to reward innovation and change that betters people's material lives. Economic growth,
in part a by-product of technological change, has long been the most sought
after public benefit from a free market.[15]

The importance of technological change in delivering desirable public
benefits from the free market suggests that law and economics should pay
serious attention to change over time. Making technological change central introduces a temporal dimension to the study of legal policy and institutions. The direction of change over time becomes important and
questions about cost must include thinking about cost over a long period
of time.

One might well want public law to stimulate innovation to better meet
public goals. But innovation and growth frequently require experimentation. And experimentation often implies failure and inefficiency.[16]
Technological change arises from the desire to increase (or avoid losing)
market share and relies on the mechanisms of competition and individual
creativity to motivate innovation. This contrasts with static efficiency,
which relies on the mechanism of calculation to match marginal production
costs to marginal consumer benefits. This matching can occur without any
innovation at all and can interfere with innovation by limiting profits available for investment.[17] In short, some tension may exist between sound
economic dynamics and perfect static efficiency.[18]

In 1999, *Time* magazine selected Jeff Bezos, founder and head of Amazon.com, as its Person of the Year. It selected Bezos because it saw him as a prescient innovator, having founded a business based on e-commerce when few could see the Internet's commercial potential. But the Person of the Year runs one of the country's least efficient businesses, in cost-benefit terms. Amazon.com usually loses hundreds of millions of dollars a year. Indeed, *Time* magazine (and the public opinion the magazine no doubt reflects) admires Bezos precisely because he is willing to lose a lot of money to pursue a grand vision. Whether or not Amazon.com ultimately becomes a successful business in cost-benefit terms, it may contribute something of value if e-commerce provides any real public benefit, such as reduced cost, better information about products and services, or increased convenience. The company itself may perish, but its leadership may change things, perhaps for the better. And if the company grows and prospers, it will be because an initially inefficient investment proved beneficial over a long period of time.

Perhaps public law should seek to encourage visionaries seeking technological change to protect the environment. Stimulating innovation requires something different from pursuit of efficiency as an ideal. Innovations come about because of an economic dynamic in the free market that supports innovation to improve our material well-being. Analyzing that economic dynamic provides a predicate for thinking about how to have public law provide a similar dynamic for innovations moving us toward public goals, such as a cleaner and healthier environment. This requires conscious use of law and economics' first insight, that economic incentives are powerful, by analyzing what incentives obtain and how they operate.

This book analyzes the economic dynamics of environmental protection and proposes analyzing regulatory programs in terms of economic dynamics rather than just static efficiency. Thinking about economic dynamics helps raise new questions about environmental protection and invites fresh thinking about the proper direction for regulatory reform. It also provides a fruitful avenue for questioning the general direction of law and economics in a variety of areas.

A brief description of economic dynamics should prove helpful at the outset. Most fundamentally, economic dynamics focuses upon change over time.

The idea of an economic dynamic analysis of law follows from some of the insights of institutional economics. In particular, institutional economics and organizational theory assume that institutions, such as government agencies and corporations, make decisions using a form of "bounded rationality."[19] Institutions' purposes and habits combine with a lack of comprehensive information to constrain the choices that institutions make. Their business and habits may make them more aware of some kinds of information and not others, and more prone to some kinds of actions and not others.

Furthermore, decision-making by institutions is "path dependent." Past actions and decisions tend to constrain the range of attractive future decisions. Thus, for example, a corporation that owns a coal-fired power plant facing a decision about how to produce more electricity will likely focus upon options that involve running its existing plant in different ways, i.e., decisions that continue along a past path. A new business deciding about how to sell electricity today may find building a new natural gas power plant attractive, for the lack of any prior commitments may make this the cheapest current option for that business.[20]

Douglass North has used these ideas of path dependence and bounded rationality to study institutional change over time. Of particular relevance to the economic dynamics of law, he focuses upon an idea of "adaptive efficiency," an aid to understanding the rules that shape economic evolution over time.[21] Adaptive efficiency concerns itself with the ability of a society to acquire knowledge, to experiment, and to creatively solve problems.[22] Under conditions of uncertainty, North claims, nobody knows the correct answer to the problems they confront, and therefore nobody knows precisely how to maximize profits.[23] Adaptive efficiency maximizes not present value, but future choice under conditions of uncertainty. It induces experiments with new methods and provides feedback mechanisms to allow for post-hoc correction of errors.[24]

These ideas of bounded rationality, path dependence, and adaptive efficiency provide valuable tools for studying the economic dynamics of law. They should prove especially valuable in evaluating prospective legislative decisions, which seek to influence a course of future events. These ideas should make it possible to better understand long-term tendencies and problems that the law should address. And they should facilitate the analysis of how law influences society and how society influences law.

An economic dynamic analysis of law moves beyond merely noticing what incentives a law provides. It would ask how the incentive provided would actually influence the people nominally affected by the incentive, using the concept of bounded rationality as a tool. This includes noticing whether the law provides an incentive that falls within the matters an institution (or an individual) will actually pay attention to, given the reality of bounded rationality. For example, the "tax on marriage" may provide an incentive not to marry, but the incentive may not influence single people's decisions about whether to marry because this factor lies outside the range of factors they will consider in making this particular choice. This implies a need for more detailed study of institutions and individual behavior.

In addition to noticing when bounded rationality makes an incentive ineffective, economic dynamic analysis of law would notice and account for non-legal incentives. Thus, even if strict standards for new pollution sources provide an incentive not to build new plants (as many scholars have argued), attributing a failure to build new plants to this incentive without observing other market factors that might influence building decisions provides an incomplete analysis. These market incentives might prove far more influential than the legal ones, overriding them in some cases or rendering them redundant in others. In other words, economic dynamic analysis should include careful accounting for what is inside and outside the bounds of "bounded rationality" to notice how law shapes (or fails to shape) society.

An economic dynamic analysis would also build on public choice analysis to better understand possible future directions for law.[25] Public choice analysis predicts that powerful interests have a disproportionate influence upon political decisions and thus upon the content of the law. Noticing whom the free market empowers and what sorts of legal rules these interests will want to pursue aids analysis of the future direction of legal rules.

An economic dynamic analysis also provides tools to understand what problems free markets will create over time. Private firms and individuals will generally consider the costs and benefits to themselves, not efficiency in the abstract, in deciding which innovations to pursue. Recognizing these bounds upon rationality provides a useful tool in predicting future trends stemming from private decisions about innovation.

Finally, an economic dynamic analysis of law should keep, rather than discard, the virtues of non-economic legal analysis. These include careful attention to legal detail and precise attention to the limits of analogies. An ideological commitment to "free market" emulation can sometimes interfere with the necessary precise analysis.

We can now generally outline the economic dynamic of environmental law that this book develops, although a full defense and elaboration will not appear until part II of this book. An economic dynamic exists that tends to diminish environmental quality over time. Any person can realize a profit by taking a natural resource and converting it into a product for sale to human beings. Hence, the free market provides a continuous incentive to find and deploy environmentally degrading innovations in order to meet human material needs and desires. Indeed, the market provides an incentive for producers to encourage expanding material desires over time, through advertising. Population increases, a natural product of fundamental biological impulses, and human desires to have more stuff accelerates this dynamic tendency to increase resource use over time.

As explained by the laws of thermodynamics, increased resource use over time diminishes the stock of useful resources that can sustain wealth. The second law of thermodynamics teaches that production converts low-entropy resources into high-entropy waste, with less economic potential.[26] Thus, over time, use of nonrenewable resources or harvesting of renewable resources at rates exceeding their ability to renew themselves should lead to reductions in wealth.

While the free market offers substantial incentives to innovate in order to create more goods, the free market offers no strong continuous incentive to innovate for the sake of improving environmental quality. The free market may encourage bigger cars that carry more people on rougher terrain, but does little to encourage the most environmentally friendly automobile possible. The free market regularly encourages entrepreneurs to take big risks in order to try and earn money satisfying our material desires, but offers no incentive for such risk taking for the sake of the environment.

The continuous possibility of profit from environmental degradation tends to limit countervailing government efforts to protect the environment. People who make profits from environment-degrading activities acquire the means to hire lawyers and lobbyists to limit government efforts to protect the environment. And all of us have an incentive to favor reduced taxation, which limits the administrative capacity of government. Over time, these efforts have a rather profound affect.

Notice that this description focuses upon the macroeconomic picture, the large shape of society over time. This description concerns itself not just with the effects this dynamic may have on any given government decision and private decision. It concerns itself with how many decisions implementing changes might occur over time. Because the free market is more decentralized than the government, many more potentially environmentally degrading private decisions will be made than countervailing government decisions. If the number of private innovations protecting the environment falls far behind the number of private decisions harming the environment, then long-term environmental degradation will prove very difficult to avoid.

This dynamic, once properly understood, should reshape our thinking about environmental law. The question of how to make each government regulation efficient becomes less important than how to address this larger dynamic. For the dynamic suggests that a set of perfectly efficient regulatory decisions will not lead to a perfectly (or even largely) efficient economy, since the number of regulatory decisions per unit of time will tend to remain small relative to the number of private decisions that do not internalize environmental costs.

This theory of economic dynamics should influence both those who believe that efficiency is the proper goal of environmental protection and those who do not. Either way, this dynamic is important.

The theory of economic dynamics not only leads to a different description of the macroeconomics of environmental law, and thereby different questions about how to achieve the normative goals of environmental law, it also provides tools for critiquing legal rules and their effects upon innovation and change. This microlevel analysis calls into question the conventional wisdom regarding regulatory design.

Economic dynamic analysis of environmental law has four uses. First, it provides a basis for critiquing the efficiency-based prescriptions for regulatory reform. Second, it should change our analysis of environmental law, reshaping our perception of what constitutes its most important characteristics. Third, it should change the questions we focus upon in thinking about reform of environmental law, inviting us to ask how to reshape the dynamics that determine long-term performance. Fourth, it improves the precision of analysis leading to new reform recommendations.

Economic dynamic analysis more generally should influence how we think about law and economics as a whole. Increasingly, legal scholars have questioned the adequacy of the traditional neoclassical efficiency-based analysis of legal regimes. The theory of economic dynamics incorporates many of the insights of institutional economics (which recognizes the limitations of institutions and the importance of change over time), ecological economics (which addresses macro-, rather than just microeconomic questions), and public choice theory (which predicts that special interests will have disproportionate influence upon decision-making). Both ecological economics and institutional economics have something to contribute to a post-Chicago School law and economics.[27] The theory of economic dynamics provides tools that can aid analysis of a variety of legal and policy questions, especially when change over time is important.

Part I explains the traditional static efficiency-based approach and uses economic dynamics to raise some questions about this approach. It includes chapters about cost-benefit analysis, free trade and the environment, and so-called economic incentive mechanisms. It questions the conventional view that standard economic incentive programs designed to meet efficiency objectives, such as emissions trading, offer a continuous incentive for innovations to improve the environment, setting the stage for a more far-reaching analysis of economic dynamics. It also clarifies the differences between economic dynamics and analysis of "dynamic efficiency."

Part II develops an economic dynamic analysis of environmental protection. It explains the value of innovation. It then compares the incentives our society creates to introduce innovations enhancing our material well-being with those it creates for innovation protecting the environment. It also compares free market decision-making structures about deploying material innovation to governmental decision-making structures affecting deployment of environmental innovation. The final chapter of part II asks whether environmental protection can keep pace with the economic dynamics driving destruction of the environment over time. Efficiency-based analysis focuses overwhelmingly on how to make good static decisions, meaning decisions that make efficient use of private sector resources spent on compliance. It has little concern with the efficiency of actual government decision-making processes, which often make government decision-making slow and uncertain. But many environmental problems come about because of numerous, relatively rapid private sector decisions, and an adequate response often requires numerous government decisions. Hence, the relative pace and scope of private and public decision-making may matter a lot in the long run.

Part III identifies three questions that an economic dynamic analysis leads to. First, should environmental law be privatized in order to improve its economic dynamics and, if so, how could this be done? Second, how can we improve governmental environmental decision-making processes to make them more efficient and fair? Third, how can we improve regulatory design to stimulate innovation and entrepreneurship? While this part presents a number of proposals in order to make the discussion more concrete, it provides only limited defenses of these proposals; this part aims to show how due regard for economic dynamics shifts the focus of debate.

Part IV reconsiders the place of efficiency-based reform in an analysis that takes economic dynamics seriously and draws out the larger implications of economic dynamics for the future of law and economics. It emphasizes that efficient use of private sector resources may conflict with efficient use of government resources. It also shows that an analysis of economic dynamics aids understanding of other legal problems often addressed through efficiency analysis. Finally, it explains that government can tailor economic dynamic reform to take into account concerns about efficient use of private sector resources, but that economic dynamic analysis may indicate a need to decrease emphasis upon cost considerations.

Economic dynamics should be central to the ongoing debate about regulatory reform. Indeed, it should play a larger role in all questions that law and economics touches. Serious attention to economic dynamics reframes analysis in productive ways. It suggests that regulatory reform should address very different questions than those that dominate current public discussion of environmental protection.

I

Static Efficiency

This part lays the groundwork for understanding how an economic dynamic approach would change the questions we ask as we evaluate environmental law and policy. It shows the efficiency-based approach's dominance. It explains that cost-benefit analysis, free trade-based limits on international environmental protection, and the most commonly discussed economic incentive measures all address the question of how to make environmental protection use private sector resources more efficiently. Since discussion of these topics dominates academic and policy discussion about reforming environmental law, discussing these topics together as manifestations of an efficiency-based approach should establish the dominance of that approach. An understanding of the role efficiency considerations play in establishing the important questions we ask of environmental policy sets the stage for parts II and III of the book, which describe how an economic dynamic approach might change analysis of environmental problems and policy.

This part also shows how focusing upon change over time raises significant questions about efficiency-based reforms and, indeed, about allocative efficiency as the dominant criterion for evaluating environmental policy. As explained in the introduction, economic dynamic analysis aims to describe the nature of change over time and what questions those changes raise about environmental policy. At this stage, however, the book will not systematically develop the idea of economic dynamics. The fuller exposition of economic dynamics begins in part II. While this part will mention some of the specific elements of the economic dynamic approach, its principal aim is simply to establish the value of the topic economic dynamics brings to the forefront, change over time, especially as a basis for critiquing efficiency-based reforms.

2

Cost-Benefit Analysis and Allocative Efficiency

Efficiency-based thinking supports the use of CBA to help determine the goals of environmental regulation. This chapter explains how increased reliance upon CBA changes traditional environmental law and reviews some of the traditional criticisms of CBA. It then develops a fresh critique from an economic dynamic perspective, thereby introducing some of the ideas that this book will develop further in part II.

Background on CBA and Allocative Efficiency

The Polluter Pays Principle and Statutes not Providing for Cost-Benefit Analysis

Most public health and environmental statutes have the goal of protecting public health and the environment, not balancing that protection against economic interests. This reflects a "polluter pays" principle that demands that the prices of goods should reflect their "prevention" costs, the full price of preventing environmental harms in the production of the goods.

At times, Congress has specified explicit numerical emission limitations for classes of polluters and has expressly listed the pollutants that agencies must regulate. But Congress more frequently leaves the task of translating general goals for environmental quality into concrete requirements for polluters to administrative agencies with authority to write rules, such as the Environmental Protection Agency (EPA), the Occupational Safety and Health Administration (OSHA), and their state counterparts.

Generally, the "polluter pays" statutes require EPA to use one of two criteria to determine the stringency of pollution control regulations. Some provisions use "health based" (or "effects based") criteria, requiring a level of reduction sufficient to protect public health and/or the environment. Others use "technology-based" criteria, requiring the agency to demand reductions achievable through available technology.

Congress usually requires or allows federal agencies and states to take cost into account in determining polluters' precise pollution control obligations.[1] Generally, the statutes make cost considerations relevant in order to meet equitable goals, such as appropriately distributing pollution control obligations among polluters. The authority to consider costs enables agencies to mandate reductions from polluters that have reasonable control options while often avoiding extraordinarily expensive controls or shutdowns. The statutory provisions requiring promulgation of health and technology based standards do not make the ratio of costs to benefits a relevant factor in agency decision-making, even when they authorize an agency to take costs into account.[2] While statutes based on technology or health-based standards have not met their overall goal of fully protecting public health and the environment, they have generated significant environmental improvement during a time of growing population, increased mobility, and economic growth.

Cost-Benefit Analysis in Environmental Law
At least two environmental statutes, the Federal Insecticide, Fungicide, and Rodenticide Act (FIFRA)[3] and the Toxic Substances Control Act (TSCA),[4] have directed the EPA to use a cost-benefit criterion to determine whether it should issue a regulation.[5] Both statutes authorize the EPA to ban the production of chemicals that enter the environment, but the EPA has rarely used these authorities, largely because of CBA.

Presidents Reagan and Clinton issued executive orders requiring CBA in all major rulemaking. As a result, agencies have long conducted CBA when they propose major regulations, even under statutes that do not authorize consideration of cost-benefit ratios. Legally, the agencies must follow the criteria set out in governing statutes, rather than a cost-benefit criterion, when they actually make a decision.[6] In practice, however, the Office of Management and Budget uses the analysis to weaken and delay environmental regulation in order to reduce regulated companies' compliance costs.[7] The 104th Congress codified the executive orders' requirements that CBA occur for major regulations in the Unfunded Mandates Act without formally displacing existing regulatory criteria.

The 104th Congress considered, but did not pass, a bill that would have forbidden enactment of regulations absent an administrative finding that the benefits of the regulation outweigh the costs. This "regulatory reform" bill and many subsequently considered bills would have radically altered the "polluter pays" statutes because they might have substituted a cost-benefit criterion for existing health and technology based statutory criteria. Some of these bills would have made findings concerning cost-benefit ratios and complex regulatory analyses judicially reviewable. At the same time, Congress eliminated the cost-benefit criterion from FIFRA, apparently recognizing that the statute had not generated adequate progress in regulating pesticides.

As this summary indicates, cost-benefit analysis has played an increasingly large role in environmental policy. Controversies about its use have occupied all of the recent Congresses, especially after the 104th Congress gave the issue so much prominence.

Allocative Efficiency and the Rationale for Cost-Benefit Analysis

Most academic CBA supporters assume that CBA, at least in theory, enhances the allocative "efficiency" of environmental regulation.[8] This section describes what this claim means.

A cost-benefit criterion (that costs should equal benefits) largely determines in theory what the environmental goal of a given regulation should be. It is a "goal-determinative" criterion.[9]

CBA does not necessarily help achieve environmental goals as cheaply as possible. Hence, it does not track a common sense meaning of efficiency as a measure of how cost-effectively one meets a given goal. If one already knows what the goal is and wants to determine the most cost-effective method of achieving that goal, there is no need to compare costs with benefits. One can simply compare the costs of various measures designed to meet the same clean-up goal and pick the least costly. CBA should not be confused with a cost-effectiveness analysis.

Economists identify CBA with achieving the "optimal" amount of pollution.[10] While many people would think that means zero pollution, economists mean something different by this phrase. Economists argue that clean air and water are amenities, just like other products we purchase on the market. To obtain these amenities, society must spend resources and forego other possible expenditures. In order to know whether one is spending the right amount on these amenities, society must make sure that it is paying a cost equal in value to the "effects" cost (sometimes called the social cost of pollution), i.e., the estimated cost of the damage the abated pollution would cause if allowed to continue.

A cost-benefit criterion may ensure that the "prevention cost" (the price of environmental controls) never exceeds the "effects cost" (the economic value assigned to the harm avoided by a proposed prevention expenditure). The notion that a polluter should pay the "effects cost" rather than the "prevention cost" comes from the theory that the government should incorporate externalities into pricing by charging a "Pigovian" tax (after the economist Pigou) equal in value to the social cost of the externality (e.g. the environmental and public health damage or effects cost) associated with production. Accordingly, the notion that a regulation might impose a cost exceeding the "effects" cost bothers those accustomed to thinking in terms of the Pigovian tax as a solution to externality problems.

CBA proponents claim that balancing costs and benefits is economically efficient using efficiency defined in the sense advanced by the economists Kaldor and Hicks.[11] Kaldor-Hicks efficiency posits that a change in a situation that creates enough wealth for a winner to fully compensate the loser is "efficient" whether or not the loser actually receives compensation. The change from a situation without pollution control to a situation with pollution control only allows citizens benefitting from the change to compensate the losing polluters if the "effects" cost (the benefit) outweighs the prevention cost (the cost). Hence, having polluters pay more than the "effects" cost is inefficient in the Kaldor-Hicks sense.

In order to treat pollution as something that has an economic cost (rather than just bad effects), one must imagine a free market where the right to a cleaner environment is bought and sold. One must imagine this because in the real world environmental and health effects have no price. If we assume that polluters have a right to pollute, then citizens, absent transaction costs, might pay polluters to reduce pollution. Economists assume that citizens would be willing to pay no more than the effects cost, which presumably reflects the value of the reduced pollution to them. Hence, charging a polluter a cost equal to the effects cost seems "efficient" because it duplicates a hypothetical free market outcome.

The polluter pays principle treats the cost of reducing pollution as a "production factor," something the polluter must pay for as part of the cost of manufacturing a product. The market for pollution control techniques determines the price, just as the market price of an essential piece of machinery must become part of the production price. CBA converts the control cost from a production factor into the price of a separate "consumer good" consisting of an environmental improvement.

The view of CBA as a means of reaching optimality has influenced policy analysts. Legal commentators, for example, have criticized current environmental law for generating large expenditures of money on trivial problems.[12] Many of the critics who claim that society spends too much on trivial harms also assert that it spends too little on serious harms, but emphasize excessive expenditures in their writing. Some of these commentators then recommend consideration of CBA as a solution to this problem.

These commentators almost always rely upon tables based upon a compilation by an economist from the Office of Management and Budget that purports to show that agency estimates of compliance expenditures per death avoided are uneven across regulatory programs. They then assert, without any supporting argument, that this table shows that regulatory programs are irrational.[13] They seem to assume that private sector compliance costs should match the benefits of avoided deaths (more or less) and therefore that the ratios of costs to benefits should be approximately even across regulatory programs.

This view that compliance costs should be approximately even across regulatory programs seems puzzling on its face. If the regulations are designed to protect the public health, then industries may well face varying compliance costs, depending on the gravity of the problems they cause and the cost of solutions. Similarly, if regulation is based on technological feasibility, then this too would tend to impose differential expense on industry.

The dominance of the neoclassical allocative efficiency concept at least explains why some academics have a normative view that costs and benefits should more or less match each other. Because allocative efficiency favors balancing costs and benefits, this concept makes an inquiry into the evenness of compliance costs across programs relevant. This inquiry seems to promise a simple answer to a potentially complicated inquiry into how well the regulatory system functions. Professor Lisa Heinzerling of

Georgetown University Law School has shown that this table contains major errors, such as listing regulations that were never promulgated and leaving out regulations that realized enormous environmental benefits.[14] Furthermore, some academics have presented this table as a portrayal of agency estimates of the value of regulatory benefits when, in fact, the table's preparer decreased agency benefit estimates in preparing the table. But some academics continue to use it, because it seems to promise a simple answer to an inquiry that the allocative efficiency concept, one with a big following, invites.

The allocative efficiency concept not only frames an analysis of traditional regulation, it then generates calls for increased reliance upon CBA as a principal component of regulatory reform. It leads at least some disinterested academics to support at least some of the industry regulatory reform agenda. Industry has long supported increased reliance upon CBA.

Scholars do not respond monolithically to the allocative efficiency framework. Justice Breyer, a leading proponent of the efficiency-based critique of regulation, recommends political insulation of regulatory policy-makers rather than CBA as a remedy to inefficiency. And those who favor increased use of CBA sometimes have varying rationales for its use.[15] Nevertheless, since support for CBA has its origins in the idea of allocative efficiency, the support for CBA shows that allocative efficiency has an enormous influence on both the analysis of existing environmental policy and on the shape of reform recommendations.

Criticisms of Allocative Efficiency and Cost-Benefit Analysis

Scholars have criticized the pursuit of "efficiency" as a policy goal and the use of CBA. Critics of Kaldor-Hicks efficiency as an appropriate policy goal point out that public decision-making should not necessarily reflect the aggregate of private preferences—the constituents of "efficiency"—even if it could.[16] First, economic efficiency is not a normative criterion. There is no obvious philosophical basis for saying that a particular economically efficient result, one representing a summation of individual preferences, is either desirable or undesirable. Second, collective decision-making should focus on a collective definition of values, rather than a summation of private preferences. Such a collective definition of goals allows for the give and take of debate, learning, changing of positions, and development and conservation of shared values, something that a summation of costs and benefits does not take into account. Even before this debate occurs, peoples' own private preferences (what they think will benefit them) often vary from their public values, what they think the society should be like. For example, a person may prefer a tax deduction on a second or third home because she believes it will benefit her, but oppose it in a collective debate about what kind of society we want, because her values suggest that society needs to feed the hungry, reduce the budget deficit, or build prisons rather than satisfy her preferences. Our view of what society should be like reflects more than the aggregate of our individual desires as consumers. It reflects a combined vision of our values and interests.

Decisions producing Kaldor-Hicks efficiency do not have the virtues associated with free market exchange. A genuine free market exchange, in theory, leaves parties to a transaction better off. This is why people freely consent to exchanges. When economists state that free market exchange is efficient, they do not mean that it is Kaldor-Hicks efficient (i.e., apt to produce losers, but also producing bigger winners), they mean that it is "Pareto optimal"—capable of making both parties better off and therefore inducing consent by all involved. Kaldor-Hicks efficient decisions lack the attractive consensual features of a free market exchange.

Allowing harmful pollution levels may injure people. CBA rationalizes allowing this when prevention costs would exceed control costs even when victims receive no compensation for the harm the pollution causes.

Critics have attacked the appropriateness and practicality of CBA because it requires one to compare two seemingly incomparable things, environmental and health effects on the one hand and pollution control costs on the other. First, because environmental and public health benefits are notoriously difficult to quantify,[17] an administrative agency will tend to undervalue them in a CBA process that requires quantification. "Soft" variables tend to get lost in the equation. Second, the government cannot and ought not assign a dollar value to human life, animal life, health, and aesthetic considerations. Third, CBA tends to devalue the benefits to future generations that stringent environmental protection offers. Fourth, benefit data to assess benefits properly simply do not exist and cannot be obtained at reasonable cost. Fifth, CBA does not take equity into account. For example, decisions to balance costs and benefits may leave those living nearest polluting facilities, often minority groups, susceptible to very large pollution burdens.

CBA in administrative proceedings poses special problems for democratic theory. CBA in theory reflects private preferences or desires. Administrative agencies, however, seem poorly positioned to understand either very well. Congress, whatever its foibles, is made up of people who have made careers out of understanding public preferences and desires. This would suggest that Congress, rather than administrative agencies, should consider CBA. This would have the added advantage of making Congress accountable for major policy decisions and confine agencies to an implementing role.[18]

An Economic Dynamic Perspective on CBA

Efficiency is a static concept. It comes from a world of real-time transactions. When you go to the grocery store to buy an apple, that apple commands a price. If you pay that price, you get that apple and the cost does not change when you walk out of the store.

CBA implicitly views regulation as a static public purchase of an environmental improvement. The public buys an environmental improvement through a decision in a regulatory proceeding. CBA assumes that the cost of making that improvement equals the polluters' compliance cost. The regulatory agency estimates this cost and predicates its decision upon whether the environmental value of the expenditure makes the compliance cost worthwhile. When it writes the regulation, the public pays the price of the regulation and gets the planned environmental improvement.

An economic dynamic perspective invites some analysis of whether the dynamics of the purchase in the regulatory proceeding resemble the purchase of the apple in the grocery store well enough for CBA to make sense. This requires enlarging our temporal perspective to see if something relevant changes when we view the regulatory proceeding as something other than a single isolated transaction.

Overestimation of Cost

The regulated party will incur compliance costs after the agency promulgates the regulation (often several years later). Studies comparing regulatory cost estimates with actual compliance costs show that regulators almost always overestimate costs.[19] This matters a lot, because the regulator pursuing optimal regulatory levels would purchase more emission reductions if the cost were lower.

Economic dynamics help explain why this occurs so regularly. Even if an agency perfectly estimated the control cost a regulation would generate prior to promulgation, the very act of enacting the regulation lowers the cost. The pre-promulgation cost estimates represent guesses based on a less-robust market than will exist after an agency promulgates a regulation. Once an agency enacts a rule, regulated companies will expect their managers to find the cheapest possible way of complying in a competitive market. If they use the technologies contemplated at the time of promulgation they will seek the lowest possible prices through competitive bidding. Furthermore, if they can find a cheaper method of meeting the regulatory target, they will use it. Hence, the equilibrium a cost-benefit criterion tries so hard to capture disappears upon promulgation of a regulation because of the economic dynamic involved.

The regulatory process creates some economic dynamics that hinder the development of accurate information about costs, even if they were predictable. Regulators rely heavily upon regulated industry for estimates of control costs.[20] Regulated industry has an incentive to exaggerate control costs in order to persuade the regulator to adopt less stringent regulation. CBA would tend to exacerbate this problem by giving cost estimates greater weight in decision-making.

Are Costs Good or Bad?

Consideration of economic dynamics invites questions about the equation of regulated parties' compliance expenditures with societal economic detriment. CBA proponents usually justify their calls for cost-benefit criteria by arguing that the societal benefits of environmental regulations should outweigh their societal costs.[21] The law and practice of CBA treat the costs that polluters pay to comply with regulatory requirements as societal costs. CBA assumes that imposition of costs upon existing polluters, whatever the environmental effects, causes a societal economic detriment. It equates the economic interests of polluters with those of society. This assumption of identity between polluter and societal economic interests produces a static approach to economic considerations.

The best justification for treating the polluter's cost as a societal cost may come from the proposition that consumers pay the price of pollution control. There are several problems with this assumption. Producers may not always be able to pass on their pollution control costs to the consumer. Sometimes market conditions make this impossible.[22] Thus, an argument that costs imposed on polluters produce a societal detriment because consumers end up paying is not always correct. The control costs may simply lessen corporate profits.

More importantly, one cannot determine whether pollution control costs constitute a "societal" detriment by focusing only on the welfare effects to the customers and constituents of regulated corporations. If pollution control expenditures produce non-environmental economic benefits that offset the cost to polluters and/or consumers (in addition to whatever environmental and public health benefits they produce), then it is hard to see why the costs are "societal."

The imposition of costs on one industry may produce economic benefits in another. Pollution intensive products often compete in the market with products and services that pollute less while providing similar services to consumers. A pollution control cost imposed on a heavy polluter may well provide an economic boon to consumers and competing producers.

Consider energy services. When we refrigerate our food or wash our clothes we use electricity and appliances. The generation of electricity causes enormous pollution problems because we rely heavily upon coal-fired power plants to generate electricity.[23] Coal competes with energy sources which cause less pollution, such as hydro power, natural gas, windmills, and fuel cells. An increase in the cost of coal-fired power generation caused by pollution control requirements may make competing sources of energy more competitive, thus increasing the revenues of producers of cleaner alternatives to coal. Consumers may have to pay higher prices for the alternative energy source than they would pay for the coal-generated energy without pollution control cost. However, this price increase may not last. The costs of competing alternative energy sources have fallen in recent years and would probably fall faster if economies of scale came into being.[24] Imposing higher costs on old "dirty" production may lead to innovations that lower the prices of products from "cleaner" competitors.

Even if the price of electricity generation remained high, consumers may still avoid those costs. The consumer's goal is to wash clothes and refrigerate food, not to consume electricity.[25] Electricity consumption is just a means to an end. The consumer may elect to purchase more efficient appliances that use less electricity. If improved energy efficiency became even more important to consumers than it is now (say because electricity prices rise) then appliance manufacturers might compete more vigorously to produce the most energy efficient product. This would eventually result in a decreased demand for electricity and a consequential fall in emissions from electricity generation. The effect may not be a cost to a consumer, but simply a cost to a producer offset by a benefit to a producer in another sector with an efficiency gain.

This example of how an increase in pollution control cost may lead to economic improvement, as well as indirect environmental benefits, is not unique. Indeed, one could demonstrate similar potential for economic improvement by imposing costs on many polluting industries that receive significant regulatory attention. The notion that increasing an industry's costs constitutes a societal economic detriment, a notion at the heart of CBA, may be wrong more often than not. Pollution control costs may prove beneficial or at least economically neutral to society.[26]

Since the world's population is growing and industrialization is spreading, world demand for products and services that meet human needs with a minimum of pollution may increase in the future. Hence, increased investment in less polluting (or non-polluting) approaches may enhance the export potential for products and services that pollute less or reduce pollution. CBA may interfere with realizing this potential.

CBA may tend to discourage investment in pollution-reducing technologies and in products that meet our needs with less pollution over time. Increased emphasis on CBA will probably create a situation where no economic actor will be able to predict whether the government will demand further environmental improvements. CBA is indeterminate in principle and creates indecision in practice. CBA may well spur more investment in lobbying and litigation and less investment in innovations to protect the environment.

Some CBA proponents cite international competitiveness concerns, which raise the specter of unemployment at home because of lack of competitiveness abroad, as a reason to favor CBA.[27] They suggest that American firms bearing environmental costs exceeding those of overseas competitors will have to reduce production or go out of business because foreign companies will be able to lower prices and steal their markets. The empirical literature suggests that environmental regulations usually produce little or no loss of competitiveness.[28] Environmental costs remain minor compared to other kinds of annual operating expenses affecting price competition for most industries.

Professor Michael Porter of the Harvard Business School has argued that appropriate stringent environmental regulation may actually enhance competitiveness by spurring innovations that may provide a competitive edge.[29] He cites examples where corporations eliminated costly materials and redesigned products to reduce costs in response to environmental regulation. Porter argues that firms can actually "benefit from properly crafted environmental regulations that are more stringent (or imposed earlier) than those" their competitors face in other countries.[30]

Professor Porter's argument forms part of a broader challenge to the traditional notion that competitive advantage derives from lowering the cost of production factors—for example, by lowering wages and environmental standards. Countries may compete by offering high wages, good education, and high environmental quality, because these things help draw talented people, spur the kinds of innovation needed for productivity growth, and increase demand for varied goods and services. The evidence shows that competitiveness tends to be positively correlated with stringent environmental regulation. Professor Porter has argued:

Detailed case studies of hundreds of industries, based in dozens of countries, reveal that internationally competitive companies are not those with the cheapest inputs or the largest scale, but those with the capacity to improve and innovate continually. . . . Competitive advantage, then, rests not on static efficiency nor on optimizing fixed constraints, but on the capacity for innovation and improvement that shift the constraints.[31]

Even for those convinced that environmental costs pose a threat to American competitiveness CBA has little to recommend it because it allows the imposition of high costs that may reduce competitiveness if enough benefits accrue. CBA may also prohibit imposing costs on manufacturers without significant foreign competition (or with ample advantages to bear significant costs without losing out) that would have beneficial environmental effects. Cost-benefit criteria have little relationship to employment or competitiveness.

Optimal Regulation as the Enemy of Optimal Pollution Levels

A cost-benefit criterion, that the cost of each regulation equal its benefits, leads to sub-optimal societal pollution levels. Environmental law typically addresses an individual pollution problem, such as urban smog, through a series of regulations demanding reductions from multiple pollution sources, because most negative environmental and health effects come from the combined impact of numerous pollution sources. An allocatively "efficient" regulatory system will not produce "optimal pollution" if it fails to address all pollution sources. The combination of a cost-benefit balanced group of regulated pollution sources and a group of sources emitting pollution that has no control costs will produce less than the optimal amount of pollution.[32] Today's statutes still leave a number of significant pollution sources, such as non-point water pollution sources, mostly unregulated. So this disjunction between optimal regulation and optimal societal pollution levels is a serious problem, even for those committed to efficiency goals.

Economic dynamic analysis—i.e., an analysis that looks at issues affecting the total number of regulatory decisions over time, not just the efficiency of each decision—shows that CBA will increase the number of unregulated pollution sources, thus exacerbating the problem. CBA has generated paralyzing transaction costs. CBA requires an extremely comprehensive and difficult analytical effort that takes enormous resources and saps agencies' abilities to comprehensively address environmental problems, which stem from numerous sources, including cumulatively significant, but small and difficult to regulate, sources. During the analytical phase, judicial review of CBA, and remand of unsatisfactory analysis (which may be very common, because non-arbitrary CBA is so difficult) pollution continues unabated. Even if the outcome of the analysis is a perfectly efficient decision, the continuation of pollution from unregulated sources that agencies never reach because of the analytic effort may well defeat efforts to have the "optimum" amount of pollution.

We have extensive experience with cost-benefit criteria under TSCA and FIFRA (prior to its recent amendment). One must examine the experience with regulatory efficiency under these statutes in order to evaluate the transaction cost problems. Both FIFRA and TSCA have enormous potential to prevent pollution. Unfortunately, the cost-benefit balancing requirements in both statutes have helped paralyze their implementation, producing not a series of finely balanced decisions, but a conspicuous failure to make decisions.

Ever since *Silent Spring*, Rachel Carson's classic book describing how accumulations of toxins in the environment threatened birds and other living creatures, the public has been concerned about the accumulation of toxic substances in the environment in general, and pesticides in particular. Yet FIFRA has generated "an analytical treadmill which makes . . . forward progress strenuous if not impossible."[33] FIFRA has required EPA to balance costs and risks before it can ban or regulate a pesticide. FIFRA required the EPA to analyze the health and environmental effects of older pesticides and to reregister those that are safe while banning those whose costs outweigh benefits. Simply analyzing the health and environmental risks (the components of benefits under CBA) has paralyzed the agency:

[T]he informational demands of risk analysis doom the regulatory process to a perpetual state of slow motion. . . . [A]fter 20 years collecting data to reevaluate the health and environmental effects of 19,000 older pesticides, EPA . . . had reregistered only 2 products.[34]

EPA Special Reviews, designed to accelerate actions against especially dangerous pesticides, have taken from nine years to seventeen years to complete since risk assessment became dominant in the mid-1970s.

When an agency does act under a cost-benefit criterion, it runs an enormous risk of reversal in court. This risk makes agencies cautious and slow, unable to move on to other problems.

This substantial risk of reversal under a cost-benefit criterion applies even to attempts to regulate clear-cut health hazards. EPA failed to phase out asbestos under TSCA after more than a decade of studying the costs and benefits of asbestos regulation. Asbestos has caused health damage so massive that tort suits have driven companies manufacturing asbestos products into bankruptcy. The cost of damages from asbestosis alone, as measured by jury awards and settlements awarded to known victims and their survivors reached $1.2 billion by 1986[35] and is expected to exceed $31.7 billion.[36] Yet the Fifth Circuit Court of Appeals reversed EPA's phase-out of asbestos under TSCA, a cost-benefit statute, in part because of reliance on unquantified benefits. The Court held that "[u]nquantified benefits can . . . permissibly tip the balance in close cases. They cannot, however, be used to effect a wholesale shift on the balance beam."[37]

This decision is not aberrational under a cost-benefit balancing approach. The federal agencies carrying out mandates to protect the public health have been able to address serious harms only because standards of judicial review have traditionally reflected Congressional desire to protect public health and the environment from harmful pollution, even when the large number of sources make quantifying the effects of any particular pollution source, and therefore any particular regulation, difficult or impossible.[38] A statute suggesting that Congress believed that EPA could precisely identify the environmental effects of particular regulations and assign them a particular weight to be balanced against costs would doom most regulations for lack of sufficient data, even where harms are serious. This would cripple modern environmental law's ability to compensate for common law's failure to adequately address serious harms having multiple sources. Ironically, when the number of pollution sources increases, the difficulty of proving that any one of them causes a particular harm cognizable at common law increases. Modern environmental law addresses this problem by authorizing regulation to prevent problems in spite of incomplete information about pollution and its effects. Cost-benefit analysis has the potential to undo this advance.

The Fifth Circuit's asbestos decision relied on a three judge plurality opinion in the Supreme Court's *Benzene* case, which creates substantial doubts about the ability of an agency to regulate potentially serious harms through a cost-benefit balancing statute.[39] The *Benzene* Court rejected a regulation limiting occupational exposure to benzene, a potent carcinogen with an unusually robust database showing serious harm to human beings.[40] The plurality interpreted parts of the Occupational Safety and Health Act as requiring OSHA to prove that target levels of exposure posed "significant" health risks in the workplace, based on substantial evidence, even though the data available to the agency only directly addressed higher levels of exposure.

Although the Supreme Court seems to have repudiated important aspects of the plurality opinion,[41] the lower courts have generally placed a substantial burden of proof on implementing agencies in its wake.[42] Statutory cost-benefit criteria in combination with a burden of proof and a substantial evidence test would force an agency to value unquantifiable benefits as trivial, no matter how serious. In order to regulate, the agency would have to have enough detailed information to be able to rationally reach specific enough conclusions about the weight of environmental effects from particular levels of pollution from a discrete group of sources to claim that these effects outweigh a particular cost. In the normal case where the science does not permit that kind of specificity, agencies will have extraordinary difficulty in coming up with non-arbitrary rationales for regulatory decisions, no matter how serious the effects contributed to by the regulated industry's emissions.

One might think that environmental statutes that only require "reasonable" decisions, i.e., rationales that are not arbitrary and capricious, would permit assigning a precautionary value to an unstudied (or poorly understood) health effect pending provision of adequate data. After all, the lack of data makes the range of reasonable, non-arbitrary actions broader, not narrower. Yet courts have imported the substantial evidence standard into judicial review even under statutes that provide for arbitrary and capricious review only, so agencies may have to provide substantial evidence in order to regulate.[43] Hence, agency culture, standards of judicial review, and judicial decisions compel agencies to give little weight to non-quantifiable benefits under a cost-benefit regime. Asbestos is a comparatively easy substance to understand. It causes a "signature" disease, asbestosis, that separates its effects on the population from that of other harmful substances. Cost-benefit requirements in TSCA have tended to paralyze implementation of the statute, as the long and unsuccessful effort to address asbestos illustrates.

The economic dynamics help explain why CBA has proven so paralyzing. The FIFRA and TSCA experience show that cost-benefit criteria create an incentive to withhold the information needed to better inform policy-makers. TSCA authorizes the EPA to regulate the thousands of chemicals that are introduced into commerce every year. EPA has the authority to require testing of chemicals to try to determine their health effects prior to their introduction into commerce. Because the EPA may ban chemicals if health effects information shows that the environmental costs of a substance outweigh its economic benefit, industry has strenuously resisted most attempts to require testing. Industry has also falsified or distorted information to hide bad health effects when they are discovered.[44] Cost-benefit criteria provide industry with additional incentives for this kind of behavior.

By contrast, EPA has been able to make some progress under some statutory provisions in the polluter pays statutes. Generally, decisions and actual progress tend to coincide with definite Congressional mandates respecting clean-up and the authority to make control decisions free from formalized risk assessment and CBA.[45]

The analytical effort that CBA demands in practice greatly slows the pace of regulation. This means that for any given unit of time, EPA will regulate fewer pollution sources than it would absent a CBA requirement. Even if consideration of CBA leads to perfect regulation, it is likely to exacerbate suboptimal society pollution levels.

Economic Dynamics and CBA: A Brief Summary

Allocative efficiency concerns tend to support regulatory reform based upon increased use of CBA. Consideration of economic dynamics leads to appreciation of problems that create doubts about the value of CBA. The economic dynamics help explain why agencies regularly overestimate regulatory cost, why high compliance cost might sometimes be good rather than bad, and why optimal regulation may actually defeat efforts to reach optimal pollution levels.

3

Free Trade and Environmental Protection

Economists usually support free trade on efficiency grounds.[1] International trade law has become a source of restraints on international and domestic environmental protection.

Trade Sanctions and National Autonomy

Increasingly, nations lack the power to protect their own people from environmental problems alone. They must cooperate with other countries to solve global environmental problems that reflect the cumulative effects of actions in many nations. Furthermore, even some domestic environmental problems, problems primarily reflecting pollution and environmental destruction in a single state, reflect the influence of global markets. Gasoline, for example, causes air pollution where burned as fuel, even if it comes from abroad. Countries need each other to solve a host of environmental problems.

The growth of environmental problems around the globe has sparked a proliferation of international environmental treaties.[2] These treaties often fail to adequately address the problem they seek to solve. A free rider problem exists. If any important contributor to an environmental problem fails to sign and implement a treaty addressing the problem, that party can often foil the efforts of all other cooperating nations, while acquiring a commercial advantage.

No international police force exists to force nations to either assume or abide by environmental commitments. Several of the most effective treaties of the 1970s and 1980s recognized this problem and employed trade restrictions to secure the cooperation of all states.

The Convention on Trade in Endangered Species (CITES) focuses on limiting trade in endangered species as a strategy. This strategy has saved several species from elimination.[3]

CITES works best when it seeks to protect a species that declines because that species (or its parts) becomes a valuable commodity in international trade. In these cases, the environmental harm, the decimation of a species, may occur in a single country. But actions in other nations help cause the decimation. For example, a poacher may kill a black rhinoceros in Kenya. He does so, however, because he can sell the rhinoceros horn. A buyer will pay a lot of money for the horn, because she can sell it in another country where people have more money and value the horn. In such a case, the purchase of rhinoceros derived products in one country contributes to the decimation of rhinoceros populations in another. In these cases, cutting off international trade may help reverse the decline of a species.

Even in cases where trade in endangered species and their parts plays an important role, CITES has not always succeeded.[4] But this is a very difficult task and very few treaties achieve consistent success in solving the environmental problems they address.

The Montreal Protocol on Substances that Deplete the Ozone Layer (Montreal Protocol) uses trade restrictions as part of a very successful treaty regime.[5] Chlorofluorocarbons and other substances used as solvents, pesticides, and refrigerants have depleted the stratospheric ozone layer that protects us from ultraviolet radiation. Increased exposure to radiation would greatly increase skin cancer and other ailments. A handful of developed countries manufactured these substances at the time of the Montreal Protocol's negotiation. While the developed countries generally recognized the seriousness of ozone depletion, each country feared that it might damage itself commercially if it limited its own production of ozone depleters, only to find other countries increasing their manufacturing. A treaty that did not bind all relevant states would not solve the environmental problem, but might involve substantial cost.

Building trade restrictions into the regime helped solve this problem. The developed countries agreed to phase out ozone depleting substances. They also agreed not to sell ozone depleters to countries that did not agree to limit their own production and consumption in accordance with the protocol. This would prevent the transfer of production to countries not party to the treaty.

This treaty has been remarkably successful in cutting production and consumption of ozone depleting substances.[6] Scientists now predict that the ozone layer, which had thinned visibly and measurably, will recover over the coming decades. This will probably prevent the occurrence of many of the expected serious impacts on human health and the environment.

Trade restrictions alone cannot explain the success of this treaty. A substantial scientific and diplomatic effort made substantial contributions.[7] The treaty also offers technological assistance to developing countries to encourage their compliance. But the threat of restrictions seems to have contributed to addressing a significant potential barrier to cooperation.

Countries have sometimes agreed to a regime of environmental protection without agreeing to employ trade sanctions for non-complying countries. Most fisheries agreements of the 1970s and 1980s, for example, did not authorize trade sanctions, but authorized each treaty party to exempt itself from particular limits on fish catches by simply giving the other parties notice that it did not intend to comply. Often, important commercial users of a fishery would give such notice (enter a reservation in the jargon of the trade), thus weakening the effects on quotas designed to conserve declining or endangered species.

In these cases, the United States frequently threatened to limit imports of fishery products from the country not complying with the international quota. These threats have often proved effective at strengthening conservation.[8] The United States and the European Union have also unilaterally required some countries to comply with standards established, not by international agreement, but unilaterally, as a condition for importing goods related to the standards. This too has strengthened conservation efforts.

Threats of unilateral actions have preceded many multilateral international environmental agreements.[9] For example, the United States' threats to unilaterally impose its own standards upon tankers visiting its ports led to an improved international agreement to limit pollution from ships.[10]

The principle of national autonomy plays an important role in international law. This principle both obstructs and facilitates environmental protection.

National autonomy may obstruct environmental protection because autonomous countries need not agree to limit their own activities contributing to a global environmental problem. Trade sanctions may ameliorate this problem. But this problem regularly poses difficulties for negotiators seeking solutions to environmental problems.

On the other hand, national autonomy means that national measures to protect a nation's own environment may apply to products coming from abroad. Hence, when the United States decided to take lead out of gasoline to protect its population from lead's neurological consequences, foreign companies selling gasoline to the United States had to eliminate lead from the gasoline they sold in the United States as well. This requirement forced foreign companies to change the way they produced gasoline on their own soil.

While this extraterritorial feature of national autonomy has been little remarked upon, it matters a lot. Without it, importation of environmentally inferior products would thwart many domestic environmental measures. This becomes increasingly true as countries seek to pursue pollution prevention strategies, rather than just end-of-the-pipe strategies. These strategies fundamentally affect the making of products. Even the threat that a domestic environmental measure might not be able to rule out its undermining through importation would make the adoption of needed environmental and public health protections less likely.

In short, trade sanctions have played an important and constructive role in securing environmental progress. The principle of national autonomy, while sometimes a problem, allows a country substantial control over environmental conditions stemming from sources consumed within its own borders, even when this effectively imposes requirements on foreign producers.

Free Trade and Efficiency

Allocative efficiency provides the normative justification for free trade.[11] Adam Smith in *The Wealth of Nations* advanced the argument that efforts to protect a country's producers by banning or levying high tariffs upon imports would not only harm the nation making the taxed or banned goods, but also the nation imposing the restriction.[12] David Ricardo refined Smith's insights into a more nuanced theory of comparative advantage.[13] The theory holds that free trade would allow each country to make that which it is best suited to make, thereby increasing worldwide production and lowering costs. This theory, with some modern refinements, suggests that free trade efficiently allocates the world's resources.

International Trade Law's Constraints upon Environmental Protection

Since the late 1940s, many countries have lowered tariffs and eliminated many import quotas under the General Agreement on Tariffs and Trade (GATT).[14] The free trade efficiency ideal sparked this agreement. Many economists believe that high trade barriers contributed to the depression preceding World War II. Contracting parties negotiated GATT in an effort to avoid repeating this disaster, relying upon the widely held view that markets would perform more efficiently with less restriction.

By the 1970s, GATT contracting parties had become increasingly interested in lowering non-tariff trade barriers. In theory, one can impede with free trade through regulation, not just through tariffs. So, this seemed like a natural extension of GATT's reach.

GATT includes a set of trade disciplines that apply to all non-tariff barriers, including environmental regulations. These disciplines require national treatment of imported goods,[15] abolition of quantitative restrictions upon trade,[16] and trade on equal terms with all GATT contracting parties.[17]

GATT's Article XX, however, generally allows countries to violate these disciplines in order to protect the environment.[18] This textual defense, however, contains some exceptions. For example, countries may not protect the environment through disguised restrictions on trade.[19]

Because of the existence of the Article XX defense and the limits of the trade disciplines themselves, GATT seemed to pose little direct threat to environmental regulation for a long time. Indeed, in the 1980s, the European Union consulted the GATT secretariat about the legality of imposing trade restrictions to help enforce the Montreal Protocol (then under negotiation) and received an informal indication that this did not offend GATT.[20] This would seem a fairly natural conclusion based on simply reading the agreement's text.

Nevertheless, in the 1980s and 1990s international trade law expanded its reach in ways that threaten environmental protection. Twice in the 1980s GATT panels (adjudicative panels of trade specialists authorized to settle disputes arising under GATT) ruled that the Marine Mammal Protection Act conflicted with GATT. In particular, refusal to import tuna caught in ways that unduly endangered dolphin violated GATT, said the panels. These decisions reflected very strained and narrow interpretation of Article XX defenses and broad interpretations of the trade disciplines themselves.[21]

While these decisions caused an outcry among environmentalists, they did not formally bind the United States. At the time, GATT required consensus adoption of panel decisions by contracting parties and this never occurred. Nevertheless, the opinions have proven informally influential.

In 1994, GATT contracting parties created the World Trade Organization (WTO) to administer GATT and several newer trade agreements. The agreement creating the WTO provides that WTO adjudicative decisions will bind parties, with or without their consent.[22] Hence, WTO panels have greater influence than the trade panels used prior to the WTO's formation.

This broad multilateral agreement to accept binding dispute settlement of a large variety of matters under very broad general principles gives the WTO an unusually strong dispute settlement mechanism. Many treaty regimes rely upon less formal means of inducing compliance or voluntary arbitration. Certainly, most international environmental agreements seem weak by comparison.

The WTO subsequently declared a United States prohibition on shrimp imports to protect endangered sea turtles contrary to GATT.[23] The United States required its own shrimpers to use turtle excluder devices. But the panel held illegal a law requiring exporters to U.S. markets to follow the same requirement.

Determining the precise scope of GATT's restraints on environmental protection remains difficult. These decisions feature very complicated and uncertain reasoning; each of these three decisions has repudiated some of the reasoning of prior decisions. Furthermore past precedent does not formally bind subsequent WTO panels. But the basic message the decisions send to national governments discourages the use of trade sanctions to enforce environmental laws.

All of these decisions address unilateral imposition of trade restrictions in at least one sense. The United States imposed the sanctions even though the international community had not agreed to use trade sanctions in any relevant multilateral agreement.

These decisions, however, cast doubt upon the legitimacy of trade restrictions agreed to in multilateral agreements. Some of the reasoning supporting the decisions (although not all) would call such uses of trade sanctions into question.[24] And the decisions left the WTO's position on this matter unclear.

Negotiators developed some of the most important international environmental agreements of the 1990s in an atmosphere casting doubt on the legitimacy of trade restrictions' use. Not surprisingly, negotiators eschewed the incorporation of trade restrictions as enforcement tools in the United Nations Framework Convention on Climate Change, the Biodiversity Convention, and several other important treaties. These agreements have accomplished very little. Indeed, the United States dealt a serious blow to the climate change regime by refusing to ratify the Kyoto Protocol to the United Nations Framework Convention on Climate Change (Kyoto Protocol), which requires, for the first time, concrete actions to address the problem. It can do so with impunity, because the climate change regime does not limit imports from non-parties in any way.

The causes of these agreements' shortcomings are complex. For one thing, they address extraordinarily difficult problems. Declines in biodiversity and climate change come about because of a huge variety of human activity. Reshaping this activity to address these problems involves an enormous challenge. But many fairly simple first steps have not been taken. It would be very difficult to establish precisely how much help trade sanctions would have given these regimes in light of the underlying problems' complexity. For example, it might be very difficult to influence United States policy through actual imposition of trade restrictions. The United States is such an economic powerhouse that economic pressures from abroad may have relatively little impact. On the other hand, one may legitimately wonder whether some threat of consequences for not ratifying the Kyoto Protocol might strengthen the hand of its domestic constituencies in ratification debates.

In any case, the WTO has cast doubt upon the legitimacy of a tool that has proven effective in solving some international environmental problems. Ironically, trade sanctions have encouraged governments to further develop GATT, to become GATT contracting parties, and to comply with its terms.[25] Indeed, unilateral trade restrictions, imposed by the United States under

Section 301 of the Trade Act of 1974, played a key role in securing approval of the agreement creating a strong WTO. Thus, panel decisions cast doubt on environmental use of tools employed successfully to strengthen the international trade regime.

The WTO has also enacted the Agreement on the Application of Sanitary and Phytosanitary Measures (SPS Agreement)[26] that effectively establishes WTO oversight of national regulation to protect public health and the environment. This authority reaches regulations that simply apply the same standards to imports that apply to domestic products in order to protect domestic public health and the environment. And the SPS Agreement applies to laws that fully comply with GATT.[27]

The new SPS Agreement, as interpreted so far by the WTO, creates hurdles for governments applying non-discriminatory, but strict, standards to protect public health.[28] Governments wishing to enact stricter standards than existing advisory international standards must base their standards on a risk assessment.[29] WTO panels have closely scrutinized national regulations to determine whether risk assessments "reasonably support" the regulatory measure at stake.[30]

Risk assessment's role has been controversial in domestic environmental regimes. Because environmental decisions must often address problems in the face of very incomplete data, the European Community tends to follow a precautionary principle. This principle suggests that environmental protection may be appropriate even in the absence of data necessary to assess a risk well. The United States has historically followed similar principles. But in recent years, industry demands for more formalized risk assessment have increased its use in the United States. Risk assessment, however, remains controversial as a matter of domestic law, because of the risk of harms from poorly understood environmental problems.

The first major case under the SPS Agreement involved European restrictions on beef from animals treated with carcinogenic growth hormones. The European Union forbids its own beef producers from using such hormones and forbids importation of beef from hormone fed animals from abroad. The WTO struck down the European Union's import restrictions under the SPS Agreement for lack of adequate scientific support.

The *Beef Hormone* Appellate Body acknowledged, in dicta, national governments' right to regulate on the basis of minority scientific views.[31] But it held that the single divergent opinion of a well respected scientist could not justify the regulatory program before it, because the scientist did not himself carry out research directly addressing hormone residues in beef fattened with hormones.[32] It also apparently held that a government cannot regulate carcinogens without scientific studies addressing the specific *application* of the carcinogen it banned, at least in the face of studies of expert opinion finding the disputed application "safe."[33] Finally, it rejected an apparently undisputed body of research identifying misapplication of growth hormones as a problem. The panel found the handful of studies on this issue "insufficient" to constitute a risk assessment of that issue.[34] This would suggest that governments cannot, under the SPS Agreement, permanently regulate any problem that has not been studied extensively, even when there is little scientific controversy about it.

The Appellate Body stated in dicta that the SPS does not require quantification of risk.[35] But its holdings, both in *Beef Hormone* and subsequent cases, cast doubt on whether any measure based on a qualitative assessment of limited information could pass muster.[36]

Article 5.7 of the SPS Agreement does authorize provisional adoption of measures on the basis of available pertinent information. No WTO panel has interpreted this language yet, because no country has defended its regulation as provisional. This language may do very little to preserve regulatory programs. Most serious deficits in scientific understanding of environmental problems last a very long time. This raises an issue as to whether authority to provisionally adopt measures on the basis of available information provides authority to keep a measure in place for a long period of time when information is lacking.

Furthermore, authority to regulate on the basis of available pertinent information might be interpreted very strictly to cripple the provision. A panel might construe pertinent information narrowly to prohibit inferences from information that only indirectly bears on the necessity for the measure. It might also interpret the requirement for a "basis" in available information to require a rather direct relationship between the information and the precise regulatory decision. When real data gaps exist, such precision will often be impossible. While it is too soon to tell how the WTO will interpret this provision, the WTO has generally interpreted language aimed at preserving measures protecting public health and the environment very narrowly.

Judicial scrutiny of scientific justifications can cripple regulatory programs where great scientific uncertainty exists.[37] Because of ethical limitations on controlled human experimentation, precise data about the effects of contaminants on human beings usually does not exist. When it does exist, it often comes into being from some very high exposure level that provides, at best, only an indirect indication of the potential seriousness of exposure to lower doses that may be more typical. In this context, burdens of proof can become critical. Whichever party bears the burden of proof in a case with very incomplete data has a good chance of losing.

The WTO has generally placed the burden of proof on regulating governments.[38] WTO panels may regularly require regulators to affirmatively prove that specific evidence directly supports their standards, rather than require complaining parties to show that regulated substances are safe or showing some defense to government inferences from incomplete data.[39] If this occurs, the SPS Agreement could significantly impede regulation, because complete data exist about very few potentially significant public health problems.[40]

This potentially intensive WTO oversight contrasts with the relatively restrained tradition of judicial review of regulation in many countries. In the United States, for example, reviewing courts recognize their own inability to appropriately evaluate complex judgments about how to proceed in the face of uncertain scientific data. They often defer to the judgments of expert agencies in such cases. This intensive WTO oversight raises questions about the WTO's competence and legitimacy.

In addition, the SPS Agreement generally requires WTO members to use the least trade restrictive means available to protect public health.[41] This least restrictive means test can significantly impede protection of the environment and public health, because it is almost always possible to imagine a less restrictive approach than the one a government adopts. Indeed, this test protects commerce from restrictions protecting public health with more vigor than the United States Constitution protects individuals from certain types of restrictions on free speech and religion.[42]

While it's too soon to tell just how much damage the SPS Agreement will inflict upon domestic health protection efforts, it seems clear that the WTO has assumed new responsibilities that limit national autonomy. The future will reveal the breadth of these incursions.

Scholars do not all agree that efficiency theory justifies all of the particulars of the WTO's new supervision of environmental protection. Nevertheless, some prominent defenses of the new regime rely heavily upon an efficiency rationale.[43] And the principle of free trade, an efficiency theory, serves as the major justification for the WTO and all of its activities.

Economic Dynamic Considerations

Thinking about several major questions that economic dynamics addresses, strategic response to incentives and macroeconomic change over time, raises significant questions about free trade. Free trade law often seems to proceed from static premises. Free trade law views all regulation having any effect on any foreign firm as a potential trade restraint. This makes sense in static terms. After all, almost any regulation that affects foreign production imposes a cost. The static theory assumes that increased cost may cause a decline in consumption. People will buy less if the product costs more.

Any tax or regulation that applies to all relevant products sold in the taxing or regulating jurisdiction may increase the cost of imports entering the taxing or regulating jurisdiction. Any tax or regulation of production processes that applies to all relevant production within the taxing or regulating jurisdiction may increase the cost of goods that the taxing or regulating jurisdiction exports. For these reasons, even-handed taxation and regulation burden international trade.

Governments may tax or regulate only domestic producers that produce only for the domestic market without burdening international trade. But a country that taxes or regulates even-handedly, i.e., that does not systematically discriminate against companies with no involvement with international trade, will often create burdens upon international trade.

This means that as international integration proceeds, even-handed regulation and taxation create more and more burdens upon international trade.[44] A jurisdiction with no international trade could even-handedly tax and regulate everything sold in the jurisdiction with no direct impact upon international trade. At the other extreme, if all sectors have some involvement with international trade, then all even-handed commercial regulation and taxation burdens international trade. More integration implies greater burdens upon international trade from routine domestic regulation and taxes. This means that a lot of regulation becomes subject to WTO scrutiny.

The economic dynamic view, a view that asks about producers' responses to regulation and taxes, calls into question the whole notion that increased cost necessarily has a negative impact on trade. As explained in the chapter on cost-benefit analysis, companies may adjust to significant added cost in ways that actually lower cost. Banning trade in ozone depleting chemicals, to take an extreme example, has led to development of cheaper substitutes. Milder regulations also often provide opportunities to find newer, better, and sometimes cheaper ways of making things.

Michael Porter's claim that strictly regulated companies often enjoy a competitive advantage also suggests that some apparently costly regulations may trigger economically beneficial adjustments, converting a negative impact upon trade into a positive or neutral one.[45] If his view is correct, then costs may matter less to international trade than previously thought. The WTO typically does not examine the actual effect of costs imposed through environmental regulations. For example, at the time the first GATT panel ruled illegal United States regulations requiring nations to refrain from setting purse seine nets on dolphins as a condition for importing tuna, the United States fleet and a number of foreign fleets had complied with this stricture for years. Yet, the WTO did not assess the impact of this measure on food consumption or tuna prices.

An understanding of economic dynamics calls into question the equation of non-discriminatory regulation and trade barriers, for non-discriminatory measures may not create inescapable burdens that matter much to the economy in the long run. This suggests that some narrowing of WTO's review may be appropriate.[46]

Herman Daly has raised a challenge to the pursuit of allocative efficiency through free trade.[47] He points out that free trade only considers the optimal allocation of fixed resources. Neoclassical economics lacks a concept of optimal scale for the economy as a whole. It lacks any notion that limits exist to the desirability of economic growth. Daly points out that if the human economy grows too much, the ecosystem upon which we depend for life will deteriorate to the point that it will fail to sustain us adequately. At some point, the marginal cost of economic growth exceeds its marginal economic benefit.

This theory of optimal scale addresses change over time, the major concern of economic dynamics analysis. It recognizes that as population and consumption increase, human beings convert natural resources into products for their own use at increasing rates. This conversion can upset the ecological system upon which human beings and other species rely for sustenance. It also raises issues of inter-generational equity.[48] This generation's consumption may limit the consumption options of the next generation.

Daly makes a distinction between economic growth and economic development. Economic growth increases the economy's consumption of natural resources and generation of waste. Hence, growth increases the "through-put" of raw materials. Economic development, on the other hand, involves more efficient use of natural capital. This means generating more economic value without increasing the use of raw materials and the generation of waste.

Daly generally favors economic development. Economic development in Daly's sense may require substantial innovation in how we meet human needs. Hence, understanding the economic dynamics applicable to innovation provides a tool useful in promoting economic development, as Daly defines the term.

Daly builds on this idea of optimal scale and the distinction between economic development and economic growth to clarify the concept of sustainable development. He suggests that sustainable development implies sustaining the maximum number of people over time with sufficient per capita wealth for a good life. He admits that this concept requires elaboration and development. But his notion of limited scale is a useful starting place for clarifying sustainable development.[49]

Daly also provides some elaboration of his idea of optimal scale. He argues that the harvest rates of renewable resources should not exceed regeneration rates.[50] Furthermore, he recommends depletion of non-renewable resources at a rate equal to the creation rate of renewable substitutes.[51] And he argues that waste generation should not exceed the assimilative capacity of the environment. While these formulations raise some questions, they do help lend some clarity to the concept of sustainable development.[52]

The preamble to GATT 1994 adopts sustainable development as a goal of the WTO. So far, this reference to sustainable development has not affected the results of WTO decisions.

The reasoning in at least one WTO decision, however, reflects some effort to take into account the WTO members' endorsement of sustainable development. In the Shrimp/Turtle case, the Appellate Body reversed a panel holding that an Article XX defense might protect some measures seeking to preserve endangered species.[53] It cited the endorsement of sustainable development as a reason to recognize that a defense for measures protecting natural resources applies to the protection of endangered species.[54] While this general affirmation did not prove sufficient to sustain the particular measure at issue in that case, the rejection of some of the panel's reasoning for striking down restrictions on shrimp caught without turtle excluder devices leaves open the possibility of the WTO upholding import prohibitions protecting endangered species in the future.

Many scholars consider the concept of sustainable development fuzzy. The phrase itself certainly does not explain what changes the concept prohibits and what sort of development it accepts. Daly's theories help clarify the concept's meaning to a degree. Without such a clarification, it's hard to make the concept of sustainable development influential in actual environmental decision-making, like that carried on at the WTO.

Daly and many environmentalists look at the ongoing destruction of species, the environmental threats to the global climate, and the impoverishing effects of environmental degradation and conclude that we have already reached the limit of sustainable growth. Many economists disagree. But this disagreement should not prevent acceptance of Daly's theoretical point, that expansion of human population and consumption of natural resources without limit will someday destroy the environment that sustains us and lead to a decrease, not an increase, in human wealth, whether or not this has already occurred. Indeed, a number of leading environmental economists agree that our resource base is finite, that the planet has limited carrying capacity, and that "economic growth is not a panacea" for diminishing environmental quality.[55]

Daly has achieved much notoriety for his opposition to economic growth and free trade. Critics should recognize, however, that Daly does not reject all increase in gross national product. He advocates a steady state economy in which "economic development" continues. One can recognize the need for limits upon economic growth and, therefore, free trade without embracing all of Daly's ideas. Surely, the concept of optimal scale, if nothing else, shows that unconstrained free trade cannot deliver benefits forever.

Indeed, one can reject Daly's specific concepts and still recognize that free trade must have at least a narrow definition, perhaps focusing exclusively upon discriminatory regulation, to function well in a world subject to resource constraints.[56] For example, economists who reject Daly's steady state approach nevertheless often recognize that optimal economic decisions must take into account ecological effects over a long period of time.[57] The bounded rationality of private individuals, who make most critical economic development decisions in our society, may not take these factors into account. Society may have to restrain trade that undermines ecological values to realize a concept of optimality that takes long-term ecological health into account, even if the optimum is not assumed to coincide with Daly's vision of a steady state.

Economic Dynamics and Free Trade: A Summary

Efficiency-based theory leads to free trade principles constraining efforts to address the many environmental problems that have international dimensions. Consideration of change over time and dynamic responses to regulation and taxes leads to some questions about the value of viewing all impositions of cost as trade restraints amenable to scrutiny under WTO administered agreements. It also leads to some questions about the value of free trade over the long term in a world where economic growth can cumulatively cause enormous environmental degradation.[58]

This chapter has also drawn a connection between economic dynamics and sustainable development. They both share a concern with change in environmental quality over time. Furthermore, Daly's theory of sustainable development suggests that government policy must stimulate innovation facilitating economic development (as opposed to economic growth increasing throughput). This highlights the importance of a major question that economic dynamic analysis helps address—how can the government better stimulate innovation protecting the environment? Chapter 11 will address the problem of translating Daly's concept of sustainable development into concrete regulatory reforms encouraging economic development as defined by Daly.

4

Cost Effectiveness and Instrument Choice

Efficiency theory also influences the choice of regulatory instruments for attaining environmental goals. The theory has a goal of selecting the instrument that achieves an environmental goal at the least possible cost. Notice that one may support efficiency-based instrument choice without supporting CBA-based goal determination. Government can select cost effective instruments to achieve goals set to protect the environment or public health or to achieve these goals to the extent possible without serious economic disruption, rather than to achieve allocative efficiency.

Efficiency theory generally demonizes "command and control" regulation as inefficient and innovation stifling, while lauding "economic incentives" as cost effective and dynamic. This dichotomy, however, rests upon important misunderstandings of regulation. At bottom, it involves an unexamined and quite problematic assumption that any program created to enhance efficiency will likewise stimulate innovation.

This chapter establishes that we need more careful thinking about how to stimulate innovation than an efficiency-based approach can supply. Since the assumption that efficiency and innovation coincide turns out to be problematic, we need a more direct analysis of how to stimulate innovation. Economic dynamic analysis provides tools for making that kind of analysis.

Traditional Regulation

This part shows that a common characterization of traditional regulation as precluding innovation rests on some fundamental errors. These errors matter to efforts to figure out how to foster innovations to protect the environment.

Environmental statutes usually encourage performance standards—a form of a standard that specifies a level of environmental performance,[1] rather than the use of a particular technique.[2] Performance standards may encourage innovation by allowing polluters to choose how to comply.[3]

Many statutory provisions severely restrict EPA's authority to specify mandatory compliance methods.[4] Several provisions require a performance standard unless EPA finds that one cannot measure emissions directly to determine compliance.[5] Even when the statutes permit work practice standards or other types of standards that *do* command specific control techniques, the statutes often require EPA to approve adequately demonstrated alternatives.[6]

This predominance of performance-based standards over command and control regulation exists regardless of the criteria used to determine the standards' stringency. Statutory provisions requiring technology-based standards, for example, instruct implementing agencies to set standards that are achievable with either existing or, in some cases, future technology. Hence, agency views concerning technological capability help determine the standards' stringency. Owners of pollution sources may generally use any adequate technology they choose to comply with the performance standards that an agency has developed through the evaluation of a reference technology.[7]

Professor Ackerman's detailed study of a particularly controversial New Source Performance Standard (NSPS) under the 1977 Clean Air Act Amendments may have indirectly contributed to frequent characterization of technology-based standards as "command and control" regulation. Economists accustomed to a static framework of analysis read Professor Ackerman's statements that this NSPS involved "forced scrubbing" as indicating that "technology-based standards identify particular equipment that must be used to comply with the regulation."[8] This NSPS, however, allowed utilities to meet their emission limitations through innovative means, although it precluded complete reliance upon techniques that could not meet the emission limitations.

This NSPS limited sulfur dioxide emissions to 1.2 pounds (lbs.) per million British thermal units (MMBTU).[9] It also required a 90% reduction from uncontrolled levels except for plants emitting less than .6 lbs./MMBTU. These cleaner plants needed only to meet a 70% reduction requirement.

Nothing in the regulation specifically required any particular technology, such as wet scrubbing. Indeed, the regulation was specifically designed to leave open opportunities for plants to meet the standards through dry scrubbing and other alternatives that were regarded as somewhat experimental.[10] Hence, if a plant operator developed some completely new approach that met these standards, the utility could use it.

Operators probably could not meet this standard solely through the use of coal washing, because coal washing, which was not a new innovation at the time, probably could not produce a 70% reduction by itself.[11] Reading Professor Ackerman's reference to the NSPS as a standard based on "full scrubbing" to indicate that the NSPS precluded subsequent innovations meeting the numerical standards would involve technical misunderstanding of the regulation. The United States Court of Appeals for the District of Columbia Circuit explained in reviewing this NSPS that *"given the present state of pollution control technology*, utilities will have to employ some form of . . . scrubbing." This necessarily implies that if utilities can develop a new technology that meets the required emission limit, nothing in the regulation precludes its use, a conclusion that necessarily flows from the numerical limits stated in the standard in any case.

This error reflects a habit of thinking in static terms. Thinking in more dynamic terms about the possibility of new technology makes it impossible to equate the NSPS Ackerman studies with specification of a technology.

In any case, one cannot draw general conclusions about the character of an entire regulatory system from a single case. A more thorough analysis of the system as a whole shows that most technology-based regulation does not command use of a particular technique. The claim that technology-based regulation does require use of the particular technology that the agency used to justify the regulation simply reflects an oft-repeated legal error, reflecting habits of thinking in static terms.

A static frame of reference has frequently led to characterization of technology-based regulation as "command and control" regulation. This term is misleading, except as applied to the relatively rare standards that actually specify techniques rather than just performance levels.

The incorrect suggestion that traditional regulation generally requires government-chosen technology would lead to a conclusion that traditional regulation legally forbids innovation. But some have made more subtle incentive-based arguments for characterizing traditional regulation as discouraging innovation.

The fundamental notion that economic incentives are powerful would suggest that polluters have substantial economic incentives to use the flexibility that performance standards offer to employ innovative means of meeting emission limitations that are less costly than traditional compliance methods. Such use of innovations saves polluters money. This incentive exists even for technology-based performance standards that did not contemplate the innovative compliance mechanism a polluter discovers.

Professor Richard B. Stewart of the New York University School of Law, however, has stated that polluters have "strong incentives to adopt the particular technology underlying" a technology-based performance standard because "its use will readily persuade regulators of compliance."[12] He does not explain why this countervailing persuasion incentive would overcome the economic incentive to realize savings through an effective and cheaper innovation, even if the persuasiveness incentive were powerful. Polluters, after all, have a number of means of persuading regulators that their innovations perform adequately if they in fact do so. First, polluters may monitor their pollution directly to demonstrate compliance. Second, in some cases polluters may eliminate regulated chemicals, which certainly demonstrates compliance.

While polluters have an equally powerful economic incentive to use cheaper alternative compliance methods for true "command and control" regulations, the polluter may have more difficulty persuading a regulator that an alternative is viable if she cannot measure emissions directly. Nevertheless, the polluter can deploy her substantial expertise to estimate the effectiveness of alternative techniques and may persuade regulators to accept alternatives. Indeed, she may persuade a regulator that a less effective technique is equally effective, because the regulator may feel insecure in second-guessing a company's judgment.

In any case, empirical studies of actual responses to regulation do not support the idea that technology-based performance standards frequently discourage innovation by dissuading companies with innovative compliant technologies from using them to meet the standards. The empirical literature actually shows that industry sometimes chooses techniques different from those relied upon in standard setting.[13]

Because so many studies claim that traditional regulation, usually described as command and control regulation, thwarts innovation, a brief review of some of the cases where this simply has not proven true seems in order. Most industry responded to OSHA's and EPA's regulation of vinyl chloride in ways that the agencies anticipated. But a proprietary "stripping process," commercialized within a year of promulgation, significantly improved polyvinylchloride resin production while lowering vinyl chloride exposure, and industry adopted a number of other innovations as well.[14] Textile manufacturers met OSHA's cotton dust standard, to a significant extent, through modernization of equipment unanticipated by the government, which was needed anyway to compete with foreign companies.[15] While a few metal foundries responded to standards for formaldehyde in the workplace through ventilation and enclosure (as expected by OSHA), most developed low-formaldehyde resins.[16] Similarly, while most established smelters responded to sulfur dioxide limits by using available technologies, copper mining firms developed a new, cleaner, process to assist their entry into the smelting business.[17] Industry responded to a ban upon ozone depleting in a wide variety of ways. In some cases, customers simply substituted soap and water for chemical solvents.[18] And operators of chloralkali plants responded to EPA regulation of mercury with some process innovations.[19] Clearly the claim that "command and control" regulation always discourages innovation is simply wrong.

Traditional regulation, however, does seem unlikely to provide *continual* incentives to reduce. Traditional regulation provides little incentive to make additional reductions once an adequate compliance cushion exists. Accordingly, traditional regulation requires repeated government decision-making in order to make progress in reducing emissions. Absent revisions of standards, progress in emissions reductions may stall. Predictable, stringent environmental regulation may provide incentives to continue reducing pollution in between regulatory decisions. A possibility of government inaction or delay (or doubts about the stringency of future limitations), however, may lead polluters to decide against making further improvements at complying pollution sources.

Critics claim that technology-based regulation involves inordinate complexity and delay. Thus, even if technology-based regulation does have the theoretical ability to stimulate continuous innovation, the complex information gathering process, the formal administrative process, and industry opposition (often backed by litigation) create uncertainty that frequently discourages continuous innovation. As we shall see, this book takes this problem very seriously.

Efficiency-Based Critique of Uniform Standards

The most persuasive theoretical criticism of traditional regulation focuses on the economic inefficiency of uniform standards—legal rules that apply the same emission limitations to all pollution sources that the rule addresses. Because facilities have unequal compliance costs, uniform standards demand relatively expensive reductions from some facilities without securing greater emissions reductions from facilities with lower compliance costs. Hence, uniform standards may use private sector resources that are devoted to pollution control inefficiently.[20]

Economists troubled by this deviation from ideal efficiency have developed recommendations for regulatory reform to solve this problem.[21] These recommendations have dominated the regulatory reform debate about instrument choice.

Cost-Effectiveness Reforms

Economists have recommended two basic reforms to improve the cost effectiveness of environmental regulation. First, economists have recommended pollution taxes. Uniform taxes would remedy the inefficiency of uniform standards. A uniform tax rate would apply to relevant emissions. Polluters with control costs exceeding the cost of the tax would presumably pay the tax and forego pollution control. Polluters who can reduce pollution for less than the cost of paying the tax would have an incentive to reduce pollution. In theory the tax would produce more reductions from those with low control costs and less from those with high control costs. This could generate, in theory, the same environmental performance for less cost than a uniform standards approach.

Economists have also recommended a system of tradeable permits. Emissions trading produces more cost effective regulation than traditional regulation to the extent that significant differences in the marginal cost of pollution control exist between pollution sources.[22] For example, if a regulator wishes to obtain 80 tons of total reductions from two pollution sources each emitting 100 tons, she could require each source to make a 40 ton reduction. If one source, Seller, has control costs of $1,000 per ton and another, Buyer, has control costs of $3,000 per ton, then this 80 ton reduction will cost $160,000 ($40,000 spent at Seller and $120,000 at Buyer). Suppose, however, that the regulator allows Seller and Buyer to trade, as long as they produce 80 tons of total reductions. Buyer may choose to pay Seller to produce 40 tons of additional reductions (beyond the 40 Seller already will produce) and forego making any reductions at Buyer. Buyer need only pay Seller a little more than $40,000 to realize a worthwhile economic benefit. Thus, in this example, pollution trading allows the same reduction for less money, $80,000[23] rather than $160,000.[24]

Clearly, emissions trading and pollution taxes reduce private sector compliance costs. Hence, they have attracted fairly widespread support among academics and policy-makers, including from some who do not support CBA.

In practice, the United States government has consistently favored emissions trading over pollution taxes. After years of EPA experimentation with limited trading programs, Congress established an emissions trading program for sulfur dioxide emissions contributing to acid rain in the 1990 Amendments to the Clean Air Act. This program requires continuous emissions monitoring, a feature generally absent from earlier programs, and has produced much better environmental results than most earlier experiments. This trading program has reduced private sector compliance expenditures.

EPA and the states have adopted numerous trading programs. Some proposals continue to be controversial, usually because of local effects that make geographical indifference questionable or concerns about adequate monitoring. The United States has even strongly supported using international environmental benefit trading to address climate change, a significant global environmental problem.

The United States has made little use of pollution taxes, even though economists usually prefer pollution taxes to emissions trading on efficiency grounds. The conventional explanation given for this preference for emissions trading comes from public choice theory. Public choice theory teaches that well organized interest groups will tend to have disproportionate influence on political decisions.[25] Since industries can be well organized special interest groups, industries' preferences will strongly influence government decision-making about regulatory instruments. Polluting industries tend to favor emissions trading over pollution taxes, because trading costs them less money than taxes would.[26] While emission taxes and tradeable permits can generate identical pollution control costs if the government has perfect information about the costs of pollution control, taxes will cost polluters more than emissions trading. Once polluters have spent the money necessary to control emissions made uneconomic by a pollution tax, they must still pay taxes on the emissions left uncontrolled because of high control costs. In a tradeable permit system, once a polluter has complied with applicable emission limitations, either through controlling its own emissions or purchasing credits from somebody else, the remaining emissions are allowed free of charge.

The government could auction off the right to pollute and bar pollution absent purchase of an allowance. This would attach a price to all pollution. But the government has not enacted requirements that polluters purchase all allowances to pollute. Hence, the emissions trading schemes actually adopted treat polluters more favorably than pollution taxes.

Economic Dynamics and Economic Incentives

While the emissions trading idea has its roots in efficiency concerns, proponents claim that it spurs innovation and provides a continuous systematic incentive to reduce pollution. This claim is important to economic dynamic analysis. Such a claim would support, at least in the realm of instrument choice, a marriage between efficiency concerns and economic dynamic concerns. This book, however, offers an alternative form of approaching the question of inducing innovation because, as we shall see, this assumption is not well justified.

Commentators rarely analyze the claim that emissions trading spurs more innovation than traditional regulation critically. Instead, they employ a conventional dichotomy that contrasts "command and control" regulation with "economic incentive" programs, including emissions trading. Because of the general prestige of the market emulation project, almost everyone simply assumes that any economic incentive program will induce continuous innovation.

I have already pointed out that the term "command and control" regulation misleads analysts about the actual operation of most traditional regulation. One might ask about the other half of the command and control/economic incentive dichotomy as well—the claim that emissions trading is an economic incentive program.

An economic dynamic perspective leads one to ask whether emissions trading is an economic incentive program. Most scholars, government officials, and practitioners equate emissions trading with economic incentives, but they do not define "economic incentives." Since economic dynamics builds on determining how various economic incentives will influence actors within the constraints of bounded rationality, it requires identification of the precise incentives of competing systems. This failure to define economic incentives leaves unsupported the suggestion that emissions trading realizes environmental goals through economic incentives, but that traditional regulations (rules that limit discharges of pollutants into the environment without allowing trading) do not. Both traditional regulation and emissions trading rely upon the threat of a monetary penalty to secure compliance with government commands setting emission limitations. Perhaps neither traditional regulation nor emissions trading should be considered economic incentive programs, because both rely upon government commands. Or perhaps both should be considered economic incentive programs, because monetary penalties provide a crucial economic incentive in both systems.

Any program to regulate or to deregulate creates economic incentives. The programs referred to as "economic incentive" programs all envision a substantial governmental role of some kind. That is why lawyers, experts in law, write about them.

Moreover, traditional environmental law creates free markets. Law performs a fundamental role in creating markets generally, and environmental law is no different. For example, laws requiring businesses to keep promises to customers and suppliers (contract) make commercial transactions possible. Laws allowing owners to forbid non-owners from using "their" property create a need for non-owners to buy or rent property from owners. Traditional environmental law creates markets, just as surely as contract and property law create markets. It establishes obligations that cause a polluter to hire people (or pay contractors) to clean up dirty facilities. This creates markets in pollution control technology, techniques, and cleaner processes, just as obligations to fulfill contractual promises and refrain from appropriating private property create markets in consumer goods.

Many scholars advocate increased reliance upon economic incentives to achieve environmental goals. But what precisely is an economic incentive? What distinguishes reliance upon economic incentives from reliance upon traditional regulation to meet environmental goals?

An economic incentive program can be defined as any program that provides an economic benefit for pollution reductions or an economic penalty for pollution. Defining economic incentives to include both positive and negative incentives includes pollution taxes in the definition. Does command and control regulation qualify as an economic incentive program under this definition? Imagine a pure command and control law. The law commands polluters to perform specific pollution reducing acts, but provides no penalties for noncompliance. This law would probably motivate little or no pollution reduction, because polluters could violate the commands without consequence. Command and control regulation only works when an enforcement mechanism exists.

Traditional regulation relies upon a negative economic incentive—a monetary penalty for noncompliance—as the principal inducement to comply with regulatory requirements, true command and control requirements, such as work practice standards, and the more common performance standards. Indeed, a traditional regulation's success depends heavily upon the adequacy of these monetary penalties.

A formal definition of an economic incentive program as any program relying on positive or negative economic inducements to secure pollution reductions plausibly applies to just about any regulatory program. To evaluate possible explanations for the dichotomy's assumption that emissions trading relies on economic incentives, but traditional regulation does not, a functional analysis is helpful. Such an analysis focuses directly on a central concern of economic dynamic analysis—how to spur innovation. A functional analysis asks whether emissions trading overcomes traditional regulation's weaknesses in spurring innovation and providing continuous incentives. This will require examination of the sources of economic inducements, the financing mechanisms, the likely responses of regulated polluters (both strategic and desired), and the governmental role in emissions trading. These questions provide the tools to develop a functional theory of economic incentives.

Because traditional regulation's dependence upon government decisions about emission limitations provides inadequate continuing incentives for innovation, the theory of economic incentives might focus on reducing reliance upon difficult government decisions. Because emissions trading depends upon government established emission limitations, it may not provide incentives for continuous environmental improvement or innovation.

A pure emissions trading model may help clarify the relationship between emissions trading, emission limitations, and incentives for *continuous* pollution reductions. Imagine a law that allows any firm that reduces pollution to trade with any firm that increases pollution but fails to mandate emission reductions from particular pollution sources. This law would accomplish little. Without regulatory limits, firms would have no obligation to make further reductions and no incentives to reduce emissions at all (or to trade).

An emissions trading program necessarily includes requirements for specific reductions from pollution sources within the trading program and allows sources to avoid the limits by trading with sources of credits. This means that some governmental body must set quantitative limits for specific pollution sources. The prospect of a financial penalty for not complying with these limits will then motivate the regulated polluters to either reduce their own emissions or pay somebody else to do so in their stead. The government set emission limitations create the demand for credits that brings a trading market into being.

Once a pollution source has complied with the underlying limits, no further incentive exists to make additional reductions. The incentive to provide reductions, either by making them at the source or purchasing credits from elsewhere, continues throughout the compliance period defined by the underlying regulations. The incentive's duration precisely matches that of a traditional regulation with the same compliance period. Once the polluters regulated by a trading program have reached an equilibrium providing the reductions that the governmental body required, no incentive for further reductions exists.[27]

The acid rain trading program provides fairly long-term incentives because it provides for staged reductions over a long period of time. However, Congress can couple long compliance times and ambitious staged reductions with either traditional regulation or emissions trading. The acid rain trading program does not provide incentives to continue reducing net emissions after an equilibrium is reached that matches the underlying reduction mandate.

Some commentators argue that emissions trading provides a continuing incentive to reduce "because the number of permits remain limited."[28] Hence, economic growth will increase the demand for permits, raise the price, and provide a continuous incentive for polluters to reduce their emissions.[29]

Limiting the number of permits does not create an incentive for continuous net emission reductions below the equilibrium level required by the program. The limit creates an incentive for permit holders to reduce emissions only to the extent that others will increase emissions. Net emissions would remain consistent with those authorized by the promulgated emission limits.

A legal rule limiting the number of permits creates incentives to avoid increases above the mandated level, whether or not the permits can be traded. The premise that a trading program limits the number of permits tacitly assumes that a legal rule prohibits the sources of additional pollution caused by economic growth from operating without purchased emission allowances. The argument that a trading program restrains growth in emissions from economic growth also requires an assumption that the trading regime imposes a cap on the mass of emissions of the sources within a trading program (as in the acid rain program). A program allowing any pollution source to operate without purchased allowances would tolerate increases in emissions associated with economic growth without demanding compensating credits.[30] Thus, even the modest argument that trading can restrain growth in emissions applies only to a particular idealized trading program, not emissions trading in general.

A traditional regulatory program that prohibits economic growth from creating additional emissions would also provide a continuing incentive to avoid net emission increases in response to economic growth. A legal rule prohibiting all non-permitted emissions would improve the environmental performance of either an emissions trading scheme or traditional regulation. Even an idealized emissions trading program does not provide a more continuous incentive for pollution reduction than a comparable traditional regulation.

One might try to save the continuous innovation theory by pointing out that once a planned reduction goal is met the government can always set another more ambitious reduction goal. If the government could be counted on to continuously revise standards then a continuous incentive to reduce would exist. But notice that this would be true whether or not the government authorized trading as the means of meeting the continuously revised goal. Even without trading, a government program that could dependably make its requirements more stringent would provide an incentive for continuous reductions. But a major critique of traditional regulation holds that it fails to provide an incentive for continuous environmental improvement, precisely because the government cannot be depended upon to strengthen standards in a predictable manner. Problems of complexity, uncertainty, and delay prevented regulators from predictably tightening limits. These problems limited traditional regulation's ability to stimulate innovation. Does emissions trading overcome this problem?

The answer seems to be no. If an administrative body sets the limits underlying a trading program, then the problems of the complexity of administrative environmental decision-making and the attendant delay may infect these decisions, just as they infect decision-making in traditional programs. The resulting uncertainty can lessen incentives to innovate, just as uncertainty about future emission limitations reduces such incentives in traditional regulation. Further, just as traditional regulation uses technological, cost-benefit, or health-based criteria to set limitations, the same criteria can be used to set the limitations governing trading programs. Also, private parties have significant incentives to litigate objectionable stringency determinations and allocative decisions.[31]

Congressional mandates of specific emission reductions may circumvent some of the problems with administrative decision-making, including hard look judicial review. Congress has, in fact, circumvented administrative problems by mandating specific cuts of named pollutants through centralized emissions trading, decentralized standard setting, and centralized standard setting. The scarcity of Congressional time may limit the frequency of

Congressional mandates. However, Congressionally set limits have often fared relatively well and should be pursued.[32] Yet the advantages of specific quantitative Congressional decision-making occur whether or not pollution sources may use trading as a means to comply with the limits.

Hence, the intuition that trading programs are easier to establish and change than traditional programs rests upon institutional confusion. Administrators establishing trading programs face many of the same problems that have interfered with efforts to make non-trading programs predictable stimulants of continuous innovation.

When one considers the possibility of strategic responses to emissions trading programs and the design complications these possibilities create, the hypothesis that administrative agencies can easily establish good trading programs becomes very dubious indeed. In addition to the usual questions that arise in a traditional regulation, such as how costly reductions will be, how much benefit they will yield, and whether they are technologically achievable, arcane disputes arise about baseline emission levels, creditable reductions, and the like in emissions trading programs.[33] Sources subject to trading have economic incentives to seek rules establishing the cheapest possible method of complying with a trading program. The cheapest methods involve claiming compliance without doing anything at all to reduce emissions. For example, a participant might seek credits for reductions that already occurred or for reductions that can occur through normal events in the business cycle, such as production declines and plant shutdowns, without accepting debits for production increases.[34] Demands to write rules that allow evasion of actual emission reduction requirements can consume regulators designing programs, increase uncertainty, and delay progress. To the extent polluters rely on reductions that would occur anyway with no legal intervention to avoid making reductions that would otherwise supplement these reductions, emissions trading simply involves evasion of pollution control.

Efforts to establish an international trading regime for greenhouse gases (pollutants that contribute to global climate change), for example, may generate fresh evasion problems. United States utilities would like to claim credit for activities abroad in order to justify avoiding potential limits on their greenhouse gas emissions at home. They may have incentives to claim credits for their role in projects that *increase* worldwide carbon dioxide

(CO_2) emissions, principally construction of new coal burning power plants. Unless the underlying emissions trading rules prohibit this explicitly, they may claim a credit representing the difference between the project built and a dirtier project that could have been built if less modern equipment was used, even though the new plant raises emissions above current levels. Of course, industry has no interest in seeking rules that assign it debits for selling equipment that raises CO_2 emissions above current levels. Debits would increase their emissions control obligations and compliance costs.

Utilities also have an economic incentive to seek credits for helping forest protection efforts abroad. If the government allows them to substitute credits for inexpensive forestry projects for more expensive pollution control efforts, they will save money. Since forests do sequester carbon that would otherwise warm the atmosphere, this seems sensible at first glance. But will the protection of any given forest have any effect on global carbon dioxide levels? If demand simply shifts to other unprotected forests, then the protection effort may not decrease net deforestation at all. Rather, the protection effort may protect one area while channeling more deforestation into areas open to logging. Hence, emissions trading may provide incentives not just to make reductions elsewhere, but to claim credits for other activities that do not have comparable value.[35]

Increased reliance on emissions trading may create a fresh incentive to resist emission limitations. Polluters may want to avoid regulation of pollution sources they own in order to protect potential sources of future credits. Thus, emissions trading may offer less of an incentive for continuous improvement (i.e., beginning before and continuing after compliance deadlines in the regulation) than comparable traditional regulation. Emissions trading, rather than providing an antidote to the problems of complex decision-making that plague traditional regulation, often provides a layer of additional complications and occasions for dispute. Furthermore, if government does not design trading programs properly, polluters will have an incentive to do nothing. This would rob these programs of whatever tendency they might have to encourage innovation.

The more difficult question is whether a well designed emissions trading program, even if it does not establish continuous incentives for innovation, establishes superior incentives to traditional regulation for some episodic innovation. Traditional regulation mandates emission reductions from specific pollution sources. Does the spatial flexibility that emissions trading offers provide superior incentives for innovation?

The trading mechanism creates additional incentives for some polluters within the trading program. It creates an economic incentive for polluters facing high marginal control costs to *increase* emissions above the otherwise applicable limit, at least to the extent that the high-cost polluters plan to purchase relatively cheap credits from other sources.[36] It also creates an incentive for polluters facing low marginal control costs to *decrease* emissions, at least to the extent the polluter plans to sell credits to sources with high costs.[37] If the market functions smoothly, then trading occurs, the incentives cancel each other out, and the net economic incentive generally mirrors that of a comparable traditional regulation.

Because a well designed trading program may induce pollution sources with low marginal control costs to go beyond regulatory limits to a greater degree than they would under a traditional regulation, commentators focusing only on the low-cost sources have argued that emissions trading creates greater incentives for technological innovation than traditional regulation. As some economists have realized, this argument ignores the incentive for high-cost sources to avoid pollution reduction activities.[38] Trading reduces the incentive for high-cost sources to apply new technology.

In theory, *emissions trading probably weakens net incentives for innovation.* If a regulation allows facilities to use trading to meet standards, the low-cost facilities tend to provide more of the total reductions than they would provide under a comparable traditional regulation. Conversely, the high-cost facilities will provide less of the total required reductions than they would have under a comparable traditional regulation. The low-cost facilities probably have a greater ability to provide reductions without substantial innovation than high-cost facilities. A high-cost facility may need to innovate to escape the high costs of routine compliance; the low-cost facility does not have this same motivation. Hence, emissions trading, by shifting reductions from high-cost to low-cost facilities, may lessen the incentives for innovation.

Some analysis of the Low Emission Vehicle (LEV) Program, a regulatory program that several states have enacted to stimulate innovation and secure emission reductions from automobiles, illustrates how emissions trading may decrease incentives for innovation. The program requires the introduction of a large number of vehicles that must meet emission standards car manufacturers can realize with fairly modest technological improvements, such as highly efficient catalysts. The program also calls for the introduction of a small number of Zero Emission Vehicles (ZEVs): most likely electric cars. The automobile industry claims that the ZEVs will be expensive to produce. One could theoretically design a program providing the same net emissions reductions as the LEV program by excluding the zero emissions mandate and authorizing a trading program to deliver the same total amount of emissions. In the short run this would theoretically produce the same emission reductions for less cost. Manufacturers would probably avoid the costly zero emissions requirement, making up the lost emission reductions by introducing more of the vehicles containing relatively modest technological improvements or by refining those improvements to get some extra reductions. However, the zero emissions provision provides the incentive to develop new technologies that may revolutionize the environmental performance of automobiles over time and even lower long-term costs. Thus, there is a tradeoff between short-term efficiency and the desire to promote technological innovation.

Another example of the way emissions trading lessens the motivation to innovate is joint implementation—an international emissions trading program proposed as a means of meeting climate change goals.[39] If the United States imposed extremely strict domestic reduction requirements upon electric utilities without allowing trading, the utilities might have to employ innovative technologies, such as fuel cells and solar energy. However, joint implementation may allow utilities to avoid these innovations. Joint implementation may allow them to claim credit for upgrading a very dirty plant abroad with off-the-shelf technology at very modest cost. These credits might substitute for relatively expensive domestic investments in innovative technologies to meet emission limitations at home.

Emissions trading advocates often cite the increased flexibility of emissions trading as a reason to expect trading to generate more innovation than comparable traditional regulation. It is unclear why increased spatial flexibility would increase innovation. Locational constraints may increase the need for innovation by requiring focused pollution control efforts that might become expensive or even impossible absent innovation. By contrast, easing the spatial constraints of traditional regulation may make it easier to choose to deploy a well understood control method at an emissions source that is cheaper to control, rather than to encourage innovation.[40]

Some analysts may regard the proposition that emissions trading creates more innovation than a comparable traditional regulation as empirically proven. The literature can easily create that impression, since so many studies claim that emissions trading creates greater incentives to innovate, using the incomplete model that does not take into account the credit purchasing sources. Furthermore, the empirical literature does claim that EPA's lead trading program and the acid rain program induced substantial innovation.[41] But I am not claiming that emissions trading never induces innovation; sometimes it does, sometimes it doesn't (as in EPA's bubble program). Rather, I am claiming that an emissions trading program will not produce more innovation than an otherwise comparable program, that is, a program using the same emission limitations without authorizing trading.

For example, EPA's lead phase-down rule, which incorporated trading, did produce a substantial change—the virtual elimination of lead from gasoline.[42] But the driver for this achievement seems to be the underlying requirement of a phase-down of lead. Faithful implementation of a traditional phase-down without trading would have produced the same change more quickly. EPA's 1985 lead trading rule supplanted a rule that required refiners to meet a standard of 1.1 grams of lead per leaded gallon, effective January 1, 1986.[43] The 1985 trading rule allowed refiners that banked purchased credits to continue exceeding these limits for almost another year, through the end of 1987.[44] EPA's 1985 trading rule actually led to increased production of leaded gasoline in 1985 (rather than purely unleaded), because the rule allowed increased production of low-lead gasoline to generate credits.[45] Furthermore, in actual implementation, inadequate reporting, compliance verification, and enforcement may have marred environmental performance, at least initially.[46] While the lead rule did lead to innovation, it's absolutely clear that an equally stringent non-trading rule would have done the same, for the performance standard underlying the rule demanded innovation.

Similarly, the acid rain trading program has produced some changes in scrubber technology, operational methods, and the use of cleaner coal, which some analysts describe as innovations.[47] But only 3 of 51 firms used inter-facility trading to emit over their allowance allocation (although 30 of the 51 did use some intra-facility averaging).[48] So, analysts should hesitate to ascribe those results to trading. Byron Swift of the Environmental Law Institute has claimed that EPA's old rate-based standards would not have permitted some of the innovations he identified, but he admits that a mass-based program without trading would have allowed all of the technologies he identified as innovations.[49]

Similarly, David Popp of the Maxwell School of Citizenship and Public Affairs finds that both the acid rain program and prior traditional regulation encouraged patenting of new technology.[50] Indeed, he finds that there was more patenting of new environmental technologies prior to the introduction of the acid rain program. He finds, however, that the programs created different types of technological incentives: the traditional program led to innovations reducing the cost of scrubbing, while the trading program produced patents improving pollution control characteristics. This very useful research, however, stops short of proving even the limited proposition that trading changes the type of innovation. For the nontrading program has much laxer limits, and a different form of limits, than the trading program. These differences, rather than the trading, may account for the observed difference.

I cite these studies because they offer some of the best and most thoughtful analysis conducted to date. But none of the studies carefully separate trading from other design elements that may influence innovation.

In any case, so far the acid rain program has not produced significant diffusion or creation of much cleaner technologies, such as natural gas power plants or renewable energy.[51] This suggests that something other than the mere existence of a trading program may be important to stimulating meaningful changes in environmental direction over time.

These observations are not meant to suggest that emissions trading is bad. Lowering short-term costs is desirable. Still, short-term savings do not necessarily coincide with the encouragement of technological advancement or long-term savings.[52] Significant up-front investment and stringent technical demands often play important roles in stimulating technological advances.

Emissions trading, traditionally considered an "economic incentive" program, may provide a less potent economic incentive to innovate than a comparable traditional regulation. An understanding of the reasons for this may contribute to a theory that would help guide design of better environmental programs. Analyzing a program's ability to provide economic incentives for pollution reduction requires an evaluation of all potentially relevant monetary flows. In simpler terms, "follow the money."

Emissions trading programs are often characterized as economic incentives because they use positive economic inducements. The lower cost source can increase revenue by reducing pollution below regulatory limits and selling credits to the higher cost source. The money to provide a positive inducement, however, must come from somewhere.

An emissions trading program produces no net incentive to do better than traditional regulation in any way because *emission increases finance emission decreases*. High-cost sources decrease costs by exceeding a regulatory limit. The savings the high-cost source realizes by exceeding a regulatory limit on pollution finance the low-cost source's "additional" pollution reductions.

Emissions trading programs, although they create no special net incentives to reduce emissions, encourage trade in emission reduction credits. These trades serve the efficiency goal admirably. The assumption that such trades serve economic dynamic goals reflects a failure to think critically about what features of a free market emissions trading imitates.

Emissions trading may provide no more incentive for continual improvement or innovation than traditional regulation. Emissions trading does not stimulate competition to maximize environmental performance. It simply authorizes some trading around of obligations the government has created. Like traditional regulation, it is only as dynamic as the government choosing the emission limitations underlying it.

Pollution Taxes

The government may tax pollution to create an economic incentive to reduce pollution. In order for a tax to encourage innovation and superior environmental performance, it must exceed the marginal costs of making additional reductions. A tax that lacks this feature creates insufficient incentives to reduce emissions below current levels.

Neoclassical economic theory supports setting tax rates equal to the "social costs" of the pollution. If a government calculation of the social costs of pollution leads to a tax less than the marginal cost of control, however, the tax will not provide an adequate economic incentive to reduce emissions. Hence, a system designed to use economic incentives to improve environmental quality must establish tax rates exceeding the marginal cost of reductions. The theory that tax rates should equal social costs assumes that environmental quality should not improve when costs of further improvements outweigh the monetary value a government body affixes to avoiding harms the pollution causes, usually mislabeled as "benefits."

This problem of criteria for determining the amount of a tax, however, flags a more general problem with taxes—some governmental body must establish the tax rate. In theory, the government can calibrate taxes to meet any given goal precisely, but a scarcity of accurate information about control costs and environmental effects makes doing so in practice difficult. Technical difficulties that thwart traditional regulation may also limit the achievements of pollution taxes.

Because a political process fixes the tax rate, taxes do not provide the escape from government decisions that the free market vision inspires. Decisions about tax rates may cause disputes. If the decision-making process involves predicting the quantity of pollution reduction a given tax will stimulate, the government must gather the same kind of information used for technology-based decision-making. The government must predict whether the tax will exceed marginal control costs at facilities in order to determine whether the tax will reduce emissions. The government must anticipate what techniques might be employed to reduce emissions in order to estimate the marginal control costs.[53] If the government wishes to establish tax rates equal to "social costs," then the government must engage in an even more difficult information gathering and analytical effort.[54] Hence, delegating authority to fix tax rates to EPA or a similar state agency might lead to delay and uncertainty comparable to that experienced under traditional regulation.

To the extent legislative bodies set rates, a less constrained process may apply, similar to that governing legislatively set emission limitations. Limits on legislative time, however, may constrain use of a legislative approach, and the legislators still might need similar information.

Because calibrating a tax to meet pollution reduction goals may prove difficult in practice, government bodies may have to revise tax rates repeatedly in order to meet public goals.[55] However, frequent revision may create uncertainties, comparable to the uncertainties traditional regulation creates, that weaken a taxation's ability to stimulate innovation. If plant operators cannot count on tax rates remaining constant or becoming stricter in a predictable fashion, they may lobby to weaken the tax system instead of implementing reductions in response to the incentive.[56]

Unlike emissions trading, a tax may offer a continuous incentive for environmental improvement. The operator can always reduce the tax by making additional innovations until the taxed pollution reaches the zero level, at least in theory. A significant tax may be necessary to secure management work on developing and implementing innovation. But the tax may provide an adequate incentive to implement further control anytime an innovation shifts the marginal cost of control to a level less than that of the tax.

Taxes may provide a greater incentive for continuous innovation than traditional regulations or emissions trading. They do not require governments to set emission levels. Like emissions trading and traditional regulation, however, they rely upon difficult government decision-making as the stimulant for emission reductions.

Efficiency, Economic Dynamics, and Dynamic Efficiency

This analysis shows that efficiency and an economic dynamic favoring significant innovation do not necessarily coincide. Emissions trading, while very efficient, does not provide a systematic continuous incentive to innovate, since all incentives for further reductions vanish when the traders reach emission limits that collectively match the government's requirements. Indeed, emissions trading may provide less of an incentive to innovate than most traditional regulation.

A pollution tax provides a continuous incentive. But it too depends upon government decision-making as a driving force. Hence, the problems of governmental timidity and delay may lessen a tax's practical impact on innovation, just as these problems limit traditional regulation's economic dynamic performance.

This lack of general coincidence between economic dynamics and efficiency highlights the need for the analysis this book provides in part II. One must focus on the economic dynamics of environmental protection directly in order to properly understand them and to produce more innovation to protect the environment.

Economists writing about the choice of regulatory instruments make extensive use of a concept that seems similar to the concept of economic dynamics, the concept of dynamic efficiency. Like the concept of economic dynamics, the dynamic efficiency concept aids analysis of the potential of regulatory instruments to foster innovation. And the analysis of dynamic efficiency overlaps with the analysis of economic dynamics in many ways.

The concepts differ, however, in the role efficiency plays. Environmental economics textbooks typically define dynamic efficiency in terms of the maximization of the present value of net benefits over a longer period of time.[57] This suggests a commitment to allocative efficiency as the correct criterion to guide environmental policy and an assumption that innovation and efficiency coincide. Current environmental law, however, does not embrace efficiency as a norm and, as I've suggested, its normative value is sharply contested. Furthermore, by framing the issue in terms of "present value" the concept of dynamic efficiency aims to maximize the welfare of present, rather than future, generations. This becomes crucial in analyzing present environmental practices that will affect the welfare of future generations, such as climate change and persistent pollutants.

By contrast, the concept of economic dynamics tries to analyze the shape and nature of future innovation in a manner agnostic to the question of efficiency for the present moment. It tries to separate the concepts of innovation inducement from the pursuit of efficiency in order to yield a clearer analysis of possible free market virtues one might emulate. This separation avoids a common confusion in the literature, the use of the term efficiency to connote two different things, the matching of marginal costs and benefits (the goal of cost-benefit analysis) and the choice of the most cost-effective means of achieving an environmental goal.[58] And current trends in economic analysis favor analysis that does not confine itself to assessing the dynamic efficiency of environmental instruments.[59] The concept of economic dynamics also does not commit itself to the welfare of this generation, but leaves open the possibility of an equal or greater commitment to the welfare of future generations or even other species.

II

The Economic Dynamics of Innovation

Part I showed how dominant the efficiency-based approach has become and how attention to change over time aids critique of that approach. It also showed that the assumption that focusing upon efficiency-based reforms produces lots of innovation is quite problematic. The analysis of particular reforms in part I illustrated the general point made in the introduction, that efficiency and creativity do not necessarily coincide.

Part II aims to establish the value of fostering innovation to protect the environment, thus furthering the case for emulating the free market virtue of fostering creativity. It provides a descriptive analysis of the economic dynamics that foster innovation aiding human comfort in a number of ways but fail to systematically foster innovations protecting the natural environment. This description identifies the features of the regulatory system that limit its ability to stimulate environmental innovation, especially radical innovation, using the free market as a positive model of how to stimulate innovation. This then leads to a set of questions in part III about how environmental law reform might better stimulate innovation.

5

Innovation's Value

Surely the existence of an economic dynamic that generates innovation, change, and economic growth raises significant questions about analysis and regulatory reform predicated upon pursuit of static economic efficiency. And, as the previous chapters suggested, pursuit of static efficiency does not always generate an economic dynamic encouraging innovation. The tension between efficiency and dynamics favoring innovation leads to inquiry about the value of innovation. The value of economic dynamic analysis depends, in part, upon the value of innovation.

This chapter claims that innovation profoundly affects both what people do and how they do it. This means that innovation over time shapes environmental problems. Innovations can make it possible to meet human needs and desires with less serious environmental impacts, by changing the products people purchase to meet their needs and the production methods used to create them. Innovation can also lower the cost of environmental protection, thus making it easier to embrace and meet ambitious goals for protection of public health and the environment. While innovation is not easy, it may be easier than limiting population or consumption, the other principal options for protecting the environment over the long term.

Defining Innovation

The literature on innovation does not agree upon a standard definition of the term. I use a definition that tries to capture what most people have in mind when they think of innovation, that is, something new.

One might define innovation as the deployment of a new way to perform a function. Producers must use various techniques to harvest raw materials and produce goods. If these techniques change, they may accomplish these tasks in a different way. Consumers use goods and services to perform various functions, such as eating, entertainment, and transportation. If the goods and services we use change because of innovation, we may change how we perform a function. Indeed, at some point, innovation may be thought of as not just changing how we perform a function, but fundamentally changing the function itself.[1]

Economists commonly distinguish innovation from invention. Innovation involves making a product or service available on the market that incorporates something new. An invention, the first development of a new product or service, does not necessarily imply an innovation, because many inventions are not commercialized.[2] While this distinction does not track common sense very well (most of us think of an invention as a type of innovation), it will prove useful for economic dynamic analysis. For this type of analysis is concerned with how the deployment of technology leads to change over time. Inventions that are not deployed have much less importance to economic dynamic analysis. This distinction also facilitates recognition of the gap between creation of a new way of doing things and its actual deployment, which is important. For, often, environmental inventions that have enormous potential are not deployed. For example, we invented electric cars, solar power, and fuel cells long ago, but they have not been deployed widely in the economy.

No bright line separates introduction of innovation from diffusion of existing technologies—the spread of a technology after its commercialization.[3] Almost all invention builds to some extent on prior practice. Federal law considers an invention worthy of a patent if it relies upon a non-obvious departure from prior art. An extension of this concept of non-obviousness helps differentiate routine technological diffusion from innovation. If an application of an old technology in a new setting is obvious, then its use represents diffusion. If an application is not obvious, then perhaps it should be considered an innovation.

Thus, for example, the use of windmills to generate electric power in California may be considered innovative (if this application was not obvious to an engineer familiar with prior applications), even though windmills have existed for hundreds of years. On the other hand, transferring technology from one use to another should not be considered innovative when the adaptation is obvious. Thus, Amazon.com may have innovated when it introduced technology allowing customers to purchase books on the Internet with a single click of a mouse. When it used the same Internet technology to sell compact disks, it did not innovate, it simply adapted the technology to a different use. The prior application of "one click technology" to books made the new application to compact disks obvious.

Types of Innovation

Environmental and Material Innovation

Professor Stewart distinguishes market innovation—innovation that firms can capture through the sale of goods and services in the market—from social innovation—innovation to meet some social goal, such as environmental protection.[4] Since this book focuses on environmental law, I will use the term "environmental innovation" to refer to those social innovations that aid the task of environmental protection.

This book defines environmental protection as the safeguarding of the natural physical environment, i.e., land, air, and water, from disruption. Of course, the protection of these media prevents harm to the living organisms that depend on this physical environment (including humans). This relatively narrow definition contrasts with a definition of the environment that embraces all that surrounds us, which would include such things as cars, houses, roads, and mixtures of residential and commercial real estate. I use this definition not to deny the importance of the man-made artifacts that surround us, but rather to sharpen and limit the issues this book addresses.

Since the free market sometimes can induce a social innovation, I will refer to an innovation that does not serve the social goal as a "material innovation" rather than a market innovation. This term also has the advantage of focusing attention upon the quality of paths of innovation[5]—a prime concern of economic dynamic analysis. An environmental innovation contrasts with an innovation that improves our material well-being, but either diminishes environmental quality or has no effect. For example, the invention of the automobile greatly improved our material well-being by making rapid long distance travel possible. But it harmed the environment, generating large amounts of air pollution and leading to the paving over of

large stretches of land. Hence, the automobile exemplifies material innovation rather than environmental innovation.

The material innovation term highlights our economy's devotion of large
amounts of resources to providing tangible material goods to individuals,
such as cars, houses, food, clothing, and toys for young and old. Application of human creativity to materials found in the environment produces
many goods for individual ownership. Indeed, even when the economy provides services, such as airplane travel, these services often involve manufacturing goods for private ownership (e.g. airlines' ownership of airplanes)
and their use to improve the material comfort of human beings only.

I do not mean to suggest that material goods serve trivial or negative values. Indeed, material goods sustain our lives in very fundamental ways.[6]
Medicines cure disease. Food keeps us alive. But material goods provide
their benefits, which involve a wide variety of values, largely to human
beings. By contrast, environmental innovation avoids rupturing the material basis for the production of goods to human beings and the lives of other
creatures. So, for example, reduced smog levels protect human health and
save trees from harm from ozone.

Qualitative and Quantitative Innovation

Often, a wide variety of products and techniques can meet human needs.
For example, a steak dinner in a restaurant, a home cooked bowl of soup,
or an appetite suppressing drug can all address the human need to eat. This
means that producers can change how they meet human needs. Indeed, they
can also, through advertising, stimulate new perceived needs and desires.

Innovation can improve our lives in several ways. It can lower the cost
of providing existing products and services. I will refer to this value as quantitative value. Innovation can meet human needs and desires that a current
product or service meets with a new, or at least improved, product or service (as in the eating example). I will refer to this as qualitative value.

The categories of quantitative (cost-reducing) innovations and qualitatively different new products and services can overlap in several ways. First,
a producer might lower cost by changing the materials in the product and
then claim, sometimes plausibly, that she has a new product. However,
some changes reducing cost, changes in production methods rather than
materials, for example, do not affect the characteristics of a product notice-

ably and some may not affect the environmental outputs of production processes either. So some quantitative innovations do not have qualitative impacts upon either the environment or the consumer. Second, one might think of sufficiently large cost reductions as in effect creating a new product. After all, if a product changes drastically in price many more people can afford it. Nevertheless, we do not ordinarily think of minor price changes as creating a new product. And if a large price change greatly changes the product's effects, it does so by making its use more frequent rather than by changing its fundamental qualities.

An example of an innovation that does not reduce dollar cost, but instead introduces qualitative improvement, may aid understanding of the distinction. The automobile did not lower our dollar outlays for transportation. We may still walk or bicycle to where we want to go and save a lot of money, especially if one considers the cost of building roads for the automobile. The automobile transformed the quality of the experience of getting from here to there. It changed physically difficult slow movement into a rapid, pleasurable experience.

Similarly, a computer with a word processing program does not offer a cheaper way to write a book than pen and paper. But it does make revision much easier. Some innovations radically change how we do things or what we do, rather than lower dollar costs.

An economist may respond to my computer example by claiming that a computer does offer a cheaper way of writing. Writing consists of two inputs—labor and materials. Computers increase the cost of the material inputs, but reduce the labor inputs, thereby reducing total costs. And an economist could make a similar argument against conceiving of the automobile as a cost raising improvement in quality.

While this desire to treat all improvements as cost reductions may be appropriate from an efficiency standpoint, it does not aid economic dynamic analysis. The very narrow view of costs inherent in the quality/quantity distinction and the focus upon changes in quality of life aid economic dynamic analysis, especially economic dynamic analysis of environmental destruction and protection.

This distinction may not serve any purpose from the standpoint of an economics that assumes that improvements in quality and reductions in cost all improve human welfare. But economic dynamic analysis takes seriously the problem of path dependence and lack of sustainability. Qualitative characteristics of products help determine the qualitative path we are on, which, in turn, limits institutional changes. In particular, the use or production of products often involves generation of pollution, so that the nature of a product and its production constitute important qualitative features. Once society chooses a particular innovation to meet important needs and desires, that choice has a qualitative effect upon the future, which can best be understood by focusing upon quality directly. This qualitative/quantitative distinction aids economic dynamic analysis of environmental law by calling attention to these qualities.

Society makes choices about how to meet human needs and desires, and even which desires and needs to meet. Simply pointing out the great variety of possible means of meeting human desires suggests that one can choose means with lesser environmental impacts. Such choices can improve environmental quality.

This description of innovation as a process of lowering cost and improving quality does not mean to suggest either that innovation is always desirable or that all economically successful innovations improve people's lives. For example, synthesis of crack cocaine (an innovation at the time) no doubt destroyed many lives. And we can raise more subtle questions about the value of legal innovations (did the introduction of television weaken our characters and capacities?).[7] But innovation can, and often does, improve our lives in various ways.

Radical and Incremental Innovation

It will prove useful to differentiate radical innovation from incremental innovation.[8] These terms bracket a continuum in the radicalness of change involved in an innovation. A radical innovation either redefines a task or changes how we accomplish an old task in some fundamental way and therefore transforms aspects of our lives. I have in mind innovations that fundamentally alter what the public gets, either in the way of productions and services or environmental quality. For example, the introduction of the automobile was a radical innovation. The car completely redefined transportation. It greatly speeded up travel, thereby making it possible to have regular face-to-face contact with friends, relatives, or business partners living hundreds of miles away.[9]

One can contrast the radical nature of the change from horsedrawn carriages to motorcars with more incremental subsequent innovations. Introducing fuel injection, while helpful, useful, and innovative, does not transform transportation itself. While it may not always be easy to differentiate radical from transformative innovations, the distinction will prove useful.

Significant innovations tend to have radical effects when they are in fairly widespread use. Hence, major innovations in general purpose technologies, technologies that many sectors of the economies use,[10] may tend to have especially radical effects.

Our society implements radical material innovation fairly frequently. Radical positive environmental innovation, however, is less common, in spite of its great potential.

A Matrix

Combining the qualitative/quantitative distinction with the environmental/material distinction yields some insights that might prove helpful in understanding the economic dynamics of environmental destruction and protection. The matrix below shows that one can have qualitative or quantitative environmental innovations and one can have qualitative or quantitative material innovations:

	Qualitative	Quantitative
Environmental	X	X
Material	X	X

A quantitative environmental innovation lowers the cost of achieving a given amount of environmental protection. A qualitative environmental innovation changes the environmental characteristics of a product or service so that it generates less environmental disruption per unit used, but without lowering the price. Indeed, some innovations may enhance environmental quality but increase the price of a product. An innovation that aids environmental protection, but does not lower the cost or improve the non-environmental quality of the good produced, may be thought of as an environmental innovation alone.

Similarly, material innovations may lower the cost or improve the quality of non-environmental products and services. In doing so they may have no effect on environmental quality or may make environmental quality worse. For example, introduction of gas guzzling sport-utility vehicles may have met consumer desires for bigger cars, but it worsened fuel economy and therefore worsened pollution problems.[11] Innovations not improving environmental quality can be thought of as material innovations alone.

An innovation that increases environmental problems will usually have slight impacts as long as few people use it. Quantitative material innovation, however, can increase the environmental impacts of products. Lowering the price of a material product can cause more people to purchase it, thus multiplying environmental impacts. For example, the introduction of the internal combustion engine was a qualitative innovation, generating a positive impact upon mobility and a negative impact upon the environment. Mass production of automobiles was a quantitative innovation, lowering the price of cars. This lowering of price multiplied sales of automobiles. This multiplied the benefits to the populace from automobile travel, but it likewise multiplied the automobile's environmental detriments. This means that efficiency improvements in material innovation, i.e., improvements that lower the cost of providing an existing consumer benefit, can exacerbate environmental impacts.

An innovation can, however, occasionally improve the environmental characteristics of a product, even if the change is aimed at improving the quality of a product for its purchaser. We might, for example, describe the introduction of a smaller car with less pollution as either an environmental or a material innovation, provided the producer made the car smaller in order to attract customers. A decision to change the characteristics of an engine to reduce pollution in ways that offer no direct non-environmental benefit to the purchaser is a qualitative environmental innovation.

Radical material innovations will probably reflect a combination of fairly important qualitative innovations with sufficient quantitative innovation to make a product's use widespread, for market innovations rely upon the price mechanism to facilitate widespread diffusion.

Describing Innovation's Value

The foregoing description and typology have already revealed something about innovation's value. Innovation can improve the quality of the products and services we buy and even offer new ways of doing things. It can also lower the price of products and services, thus making them easier to buy. Hence, as pointed out in the book's introduction, innovation plays a major role in producing economic growth.

Environmental innovation can reduce the cost of achieving environmental goals. It can also make possible better environmental quality than current technology allows. General purpose technologies, in particular, can have a profound impact upon the physical environment.

The discussion also implies that environmental innovation can take the form of changing the products consumers buy or changing the production processes that make existing products. It remains to say a little more about the relationship between innovation and efficiency and to highlight the role of radical innovation.

Lowering Cost Inefficiently

Innovation may reduce the cost of providing a given product or service.[12] Often, however, a material innovation that may reduce the cost of meeting a goal over a long period of time may cost more than standard techniques in the short run. Development of a new technique frequently requires large investments of capital and human resources.

Innovations in how we do things may lower the cost of providing environmental protection. Thus, for example, some companies have responded to costly mandates to treat hazardous wastes by re-engineering production processes to avoid use of the hazardous chemical inputs triggering the treatment requirement.[13] A treatment technology and the redesign of the production process provide two contrasting means of meeting an environmental goal. The process innovation may lower the cost of environmental improvement.

Successful environmental innovation, like other kinds of innovation, may have high initial costs. A numerical example for an environmental innovation may help. Suppose that an end-of-pipe control limiting air pollution at a chemical plant requires an initial capital outlay of $100,000 and $5,000 a year in operational costs thereafter. Suppose that one could re-engineer the process making a specific chemical for $150,000. This cost includes the costs of paying employees (and perhaps consultants) to figure out how to change the manufacturing process and some capital costs in changing the configuration of the plant. Suppose that this re-engineering cut back on the use of raw materials, generating a $1,000 a year cost savings.

The cumulative cost declines over time for the re-engineering solution, while the cumulative cost for the end-of-pipe control increases. Assuming a discount rate of zero, over a period of ten years or more the innovative solution saves money (see table 5.1). From a short-term perspective, however, the end-of-pipe solution appears cheaper. Over a period of nine years or less, the end-of-pipe solution saves money. Application of a discount factor would make the year in which the re-engineering solution started to become economic later. But it would remain true that the innovative re-engineering solution is more expensive in the near term, but cheaper in the long run.

This implies that cost has a temporal dimension. One cannot address questions about the relative costs of solutions to problems without specifying the time periods involved. To the extent that efficiency considerations focus regulators on short-term costs, consideration of efficiency may be economically counterproductive in the long term.

Expensive innovation, from a short-term perspective, may be economically desirable, from a long-term perspective. In the long run, what appears

Table 5.1
Cumulative and Incremental Costs Comparison Over Time

	Year 1	Year 2	Year 3	Year 4	Year 5
Re-engineering					
Incremental cost	$150,000	$(1,000)	$(1,000)	$(1,000)	$(1,000)
Cumulative cost	$150,000	$149,000	$149,000	$147,000	$146,000
End-of-pipe					
Incremental cost	$100,000	$5,000	$5,000	$5,000	$5,000
Cumulative cost	$100,000	$105,000	$110,000	$115,000	$120,000
	Year 6	Year 7	Year 8	Year 9	Year 10
Re-engineering					
Incremental cost	$(1,000)	$(1,000)	$(1,000)	$(1,000)	$(1,000)
Cumulative cost	$145,000	$144,000	$143,000	$142,000	$141,000
End-of-pipe					
Incremental cost	$5,000	$5,000	$5,000	$5,000	$5,000
Cumulative cost	$125,000	$130,000	$135,000	$140,000	$145,000

to be expensive innovation by today's standards may actually be more economical, as it might lower overall costs of meeting environmental goals.

Hence, innovation sometimes serves a quantitative value, reducing the cost of carrying out a task we are already pursuing. Even when innovation serves this goal, it may do so inefficiently through a series of initially unprofitable investments.

Radical Qualitative Innovation

If a technique is really new, it may have the potential to radically change both the inputs and the environmental characteristics of a set of activities. Conversion to an economy based on fossil fuel, for example, has produced a wide variety of negative environmental impacts. Selection of such "general purpose technologies" as the internal combustion engine "may have profound impacts" upon "the long-term environmental prognosis."[14]

Cars lead to ecosystem destruction, oil spills, and massive amounts of routine pollution in all media. The relevant pollution problems include tailpipe emissions, emissions during car manufacturing, air and water pollution from petroleum refining, and oil spills. Cars also kill tens of thousands of people every year directly, through car accidents.[15] Similarly, coal-fired power plants produce enormous quantities of air pollution[16] and also require environmentally destructive coal mining.

Conversely, radical qualitative environmental innovation can produce a much better environmental future. For example, a change from coal-fired power plants and gasoline-powered cars to energy derived from fuel cells would make an enormous difference in environmental quality. Coal-fired power plants and vehicles generate enormous amounts of sulfur dioxide, nitrogen oxides, carbon dioxide, and toxic pollutants.[17] This pollution contributes to climate change, acidifies ecosystems (killing fish, trees, and other life), helps cause serious respiratory problems that afflict millions of people, and increases risk of cancer and reproductive toxicity. Even drastic reductions in these pollutants through controls and burner efficiency from uncontrolled levels leaves substantial amounts of pollution. Basically, burning coal and gasoline is a very dirty business.

The federal government has spent thirty years gradually improving the control of coal-fired power plants,[18] rather rapidly reducing emissions from cars,[19] limiting oil refineries' pollution, and addressing the environmental effects of coal mining, oil drilling, and oil transportation. After three decades of regulation and millions of dollars of control costs, these industries still play an enormous role in damaging public health and the environment.

The fuel cell would introduce an enormous qualitative improvement in people's lives. It would produce no appreciable pollution.[20] Using fuel cells produces none of the environmental impacts associated with coal and drastically reduces most of the emissions associated with the production of gasoline. Introduction of fuel cells would be a radical innovation environmentally. It would fundamentally change how we produce electricity. Widespread use of fuel cells would radically improve environmental quality, permitting dead lakes to recover, drastically lowering urban smog levels, greatly ameliorating the threat of global warming, and saving lakes, streams, and mountaintops from destruction during mining operations. Since a single fuel cell in a home can supply all of that home's power needs for a period of time, one might not need large, unsightly wires that now run everywhere to deliver electricity to houses. This would be a radical change.

Some prominent legal scholars, perhaps under the influence of the neoclassical framework's relentless focus upon efficiency, have often failed to fully appreciate this qualitative dimension of innovation in the environmental context.[21] They tend to write about environmental innovation as a way of reducing the short-term cost of meeting a fairly narrow, centrally planned goal and assume that innovation has no value if it does not do that. They assume that environmental innovation that does not reduce the cost of achieving a discrete environmental goal is worthless "innovation for innovation's sake."

Qualitative environmental innovations, especially radical innovations changing general purpose technologies, may make seemingly unreachable goals attainable. Scholars too often employ a static framework that involves political selection of pollution control goals based on some static criterion and the pursuit of the cheapest short-term means of achieving the goal. This analytical framework yields some very useful insights, but it obscures an important relationship between goals and means. Our fears about the cost of achieving environmental goals limit the goals we set and often contribute to failures to meet the goals we do establish. And falling costs of initially expensive innovations over time can change the parameters, making it much easier to set and meet ambitious goals.[22]

Furthermore, since innovations frequently lay the groundwork for further innovations and further improvements, promotion of innovations has value beyond the immediate cost savings an innovation might offer. Economists recognize that knowledge has "an intertemporal spillover effect." This implies that once a firm makes a substantial innovation, the knowledge gained in doing that might help bring about further improvements. Indeed, intertemporal spillover effects may make a very expensive technique cheaper than seemingly less expensive existing techniques as manufacturers learn to improve the newer evolving technique.

The examples given below suggest that radical environmental innovation may have powerful positive effects because of such innovation's capacity to address multiple problems. A regulatory system seeking to improve social performance without disrupting existing industries must deal piecemeal with each problem a technological pattern produces. This explains why several different laws regulate the symptoms of fossil fuel use, oil spills, refinery emissions and discharges, mining and drilling waste, traffic congestion, car emissions, and power plant emissions, for example. Radical environmental innovation has the potential to address, or even eliminate, all of these problems at once, for radical innovations can disrupt environmentally destructive technological patterns with many components.

Not all environmentally radical innovations involve unalloyed environmental improvement. Nuclear power, for example, drastically reduces air pollution. But it creates a by-product, radioactive waste. Disposing of that waste poses great challenges, since such wastes remain potentially dangerous for thousands of years. Nuclear power also involves risks of sudden catastrophes. Nuclear power, however, when it appeared in the middle of the last century, surely was an environmentally radical innovation. It involved a radically different technology with a radically different set of environmental characteristics and problems. Hence, some radical innovations may substitute new problems for old ones.

Radical innovations may change the nature of inputs into a process, thereby greatly changing the nature of environmental disruption involved in providing raw materials for a process. It may also radically change the nature of the pollution the process generates. It may affect several pollutants at once. It may eliminate certain types of pollution altogether.

Path Dependence, Lock-in and Sustainable Development: How Innovations Shape the Future

Technological choices play a significant role in determining the future's shape. A look back at the history of the automobile helps give a sense of this. Realization of the automobile's promise required large scale manufacture and distribution of gasoline. It also required creation of a vast infrastructure of roads supporting the automobile. This infrastructure helped eliminate the need for people to live close to their workplaces, fueling the growth of suburbs and contributing to urban decline. The automobile reconfigured the landscape of the United States.

These changes produced path dependence, because the automobile changed what we expect from daily life. We now expect to be able to live many miles from where we work, to have paved roads in almost any place we wish to visit, to visit friends who live hundreds or even thousands of miles away, and to travel alone or with people of our choosing most of the time. Cars have transformed our culture.

People's expectations about their lives profoundly affect the demands they make for products and services. These demands shape environmental problems and limit potential solutions.

This illustrates a kind of path dependence for individuals. We who have grown up with the automobile use and expect rapid, individualized transportation. We have developed habits and lifestyles built around this expectation. Relatively few among us will tolerate changes, however beneficial in the abstract, that unsettle these expectations.

Further, entire industries have grown up to supply cars, gasoline, and roads. These industries consist of specialists in the skills and techniques needed to make these tools. These institutions probably possess great ability to adopt innovations that refine these technologies, but may be disinclined to develop radical alternatives to these technologies.

These institutions and the investments they have sunk into chosen technologies may make innovation on a very different technological path difficult, in effect locking in technologies that might be socially undesirable in some ways.[23] For example, we currently seem locked-in to technologies heavily dependent on fossil fuels, in spite of their major role in creating an enormous number of environmental problems.[24]

Future technological choices will have similar impacts. The technologies we choose in the future will create a set of expectations and demands that profoundly affect the environment we live in, what we see every day, what we experience, the shape of the landscape, and the health of our ecosystems and our children. In short, the direction of technological change is profoundly important.

Environmental economists assign environmental technological innovation a central role in realizing the goal of sustainable development. The chapter on free trade already explained that Herman Daly has identified limits on increased throughput as central to realizing sustainable development. In theory, we can realize sustainable development through global declines in the amount of per capita consumption or reductions in population. If each of us consumes less there will be less pollution over time and less resource depletion. Similarly, if we reduce human populations, we will consume less and have the same effect. But limiting consumption would prove cruel if applied to the poor of the world, who may not have enough food and clothing now for a decent life. If applied to the relatively rich, consumption limits would still prove extremely difficult to apply in practice. People do not gladly embrace declines in their standard of living. Limiting population growth, rooted as it is in basic biological instincts, also poses daunting challenges. Absent population or consumption decreases, limits on throughput imply meeting current human consumption goals with less resources. This necessarily implies a process of innovation to provide goods and services with less use of natural resources as inputs and less pollution outputs. In the long run, the supply of material inputs from the earth and earth's ability to absorb waste are limited, but our capacity to use and increase knowledge for the sake of environmental innovation has less finite limits.[25] While technological innovation, especially radical environmental innovation, may prove difficult, the alternative (or more precisely, complementary) paths toward sustainable development are even more difficult.

Caveat: Technology Is Not Everything

Technological innovation will probably not solve all of our environmental problems, even if it may ameliorate or cure a great many of them. Consider, for example, the mass extinction of species we are currently experiencing.[26]

This decline in biological diversity seems impervious to simple technological fixes. Much of the species loss occurs in tropical rainforests. While some of the destruction of rainforests might be curbed by technological improvements, much of it remains fundamentally linked with deeper socio-economic and demographic problems. Companies sometimes destroy rainforests to get at resources that sell on international markets. Perhaps substitutes for some of these resources would ease these pressures. But peasants in desperate need of land to grow food have also burnt large stretches of rainforests.[27] It's difficult to imagine a simple technological fix to such a complex problem.

One may question the importance of technology, even where it plays a fairly central role. We, after all, choose how we live. Technologies may help shape human preferences, but they also reflect them. These preferences reflect our views about what matters. If we really value the earth, we will not choose technologies that tend to destroy it. Hence, the heart of the matter may be ethics, culture, and human conduct, not technology.

Although this book takes the view that technological change is important, it does not deny the importance of ethics, culture, and individual choice. The automobile did not reconfigure the landscape by itself. Human beings decided to drive, to vote for politicians who favored road building, and to move to the suburbs. Human beings can decide to use mass transit, to live in communities that do not depend on automobiles, and to live close to where they work.

Economic dynamics and technology, however, play a role in shaping actual behavior. Companies manufacture cars to make money and provide a convenience to each purchaser. Once enough people become accustomed to enjoying this convenience, it becomes part of the culture and helps shape expectations. Soon, drive-in restaurants and drive-by bank teller machines become part of our culture. And the politics of devoting funds that might be used for highway construction to fund mass transit or bicycle paths becomes quite treacherous. Walking becomes almost impossible in some places and less pleasant in many areas, as anyone who ever tried to walk along a major parkway or interstate lacking sidewalks or cross a busy street in an urban center knows. Technological success in transforming a culture is not inevitable, and human beings sometimes successfully resist or channel technological change for ethical reasons. But the decentralized economic dynamics that allow companies to profit from each unit sale of a product, before the implications for the wider culture become clear, make technology a powerful influence.

Economic dynamic analysis helps explain how technology gets chosen, by identifying the relevant incentives and dynamics. It focuses on these dynamics rather than ethics and culture, because government probably can affect these dynamics more easily than it can change ethics and culture. This book's focus on law and policy implies no rejection of the importance of ethics and culture.

Technology then has both qualitative and quantitative value. The quantitative value involves lowering the cost of accomplishing old tasks. Qualitatively technology changes how we live and what we experience. Radical qualitative environmental innovation would have special value for meeting environmental goals. In the next chapter we begin exploration of the issue of whether adequate incentives exist for environmental innovation.

6

Economic Incentives for Innovation

This chapter compares the economic incentives encouraging innovations improving our material well-being (material innovation) with the economic incentives encouraging innovations protecting the environment (environmental innovation). This comparative analysis of economic dynamics will help clarify the importance of a central challenge that environmental law and policy must face—how to appropriately influence the shape of future innovation.

Economic Incentives for Material Innovation

Generally, our economy provides a continuous and fairly systematic incentive to develop and apply material innovations. A person or firm that develops a new way of doing things that produces a material improvement may make money through the innovation and therefore has some incentive to employ useful innovations.[1] This incentive remains continuous. At any time, an innovative material improvement may find a market.[2]

The economy provides a positive incentive to innovate by creating a prospect for economic gain. It also provides a negative incentive, however. Firms that do not innovate may lose market share to innovative competitors.[3] Indeed, they can go bankrupt by failing to keep up with competitors. Hence, both greed and fear help motivate many worthwhile innovations.

The incentive to innovate extends to inefficient technological change, at least in the short term. Amazon.com presents an example of the power of the incentive for inefficient technological change. Jeff Bezos, the company's founder and CEO, has made a large fortune by using the Internet to lose hundreds of millions of dollars. Of course, those who finance Bezos, venture capitalists and shareholders, did so in the hope that the company would one day earn a profit, as it eventually did. But they financed Amazon.com for many years, even when the company earned no profits. Furthermore, once Amazon.com began selling books on the Internet, its competitors began to do so as well. Fear of losing market share motivated this.

While the Amazon.com example may seem extreme, it actually exemplifies a feature of the free market that excites widespread admiration—its stimulation of risky behavior by those who believe in an innovation's economic potential. Many individuals risk their financial lives to bring new techniques and inventions to markets. And even existing firms sometimes willingly invest in ideas that do not generate money in the short run in order to make more in the long run. American firms sometimes endure criticism for not doing this enough. Too much focus on the short-term economics of transactions has often been viewed as a problem likely to cause a firm to suffer losses in the long term and to harm the national economy.

Indeed, a large number of leading companies that are generally profitable and deliver valuable goods and services to the public began as extremely inefficient experiments. Consider FedEx Corporation. In its first years, it lost a lot of money building the infrastructure necessary to implement reliable overnight mail delivery. Once it had put the necessary infrastructure in place, however, it began to profitably provide overnight mail service.

Many economists who study innovation recognize that viewing the decision about how and whether to innovate as the product of a cost-benefit analysis is problematic. For some innovations, especially radical ones, the initial costs may be high and the future costs (or benefits) unknown. Economists recognize that substantial technical, market, and general business uncertainties often make accurate prediction of costs and benefits from research and development (R & D) expenditures impossible.[4] Firms pursuing radical innovations may experience large and unpredictable losses and failures. For example, early research into Herbert Dow's process for producing bleach ended in 1893, when an explosion destroyed Dow's laboratory at Midland Chemical Company.[5] Dow Chemical's efforts to develop this invention a few years later triggered more production-disrupting explosions and red ink for several years, until the company began using paraffin purchased at the local drugstore. Clearly, Dow Chemical could not have anticipated these setbacks, or the source of its ultimate success. And it is a good thing that it did not, because its persistence eventually paid off, bringing a profitable product to market that has generated large sales over a long period of time. Our society tends to glorify taking risks to introduce innovative improvements, because some of our most wonderful achievements involved adjustments to early failures.[6] Willingness to undertake large risks for potentially large future payoffs characterizes one of the most widely and properly admired features of our capitalist system.

Because the numerous uncertainties surrounding decisions about investments aimed at creating innovations make "optimization" difficult, some economists use models that do not presume an a priori calculation of efficiency to describe the innovation process.[7] Herbert Simon's idea that institutions use "bounded rationality" to make "satisficing," rather than efficient, decisions has influenced many students of innovation.[8] The "evolutionary model" based on this idea assumes that firms use "routines" and "rules of thumb" to make decisions about R & D investments.[9] Even those economists who use an "investment subject to market failure" approach, which posits that firms plan R & D through calculations about future revenues and costs, recognize that the uncertainty of outcomes from innovation is very significant.[10]

Importantly, economists frequently claim that the uncertainty makes financing R & D very difficult.[11] Furthermore, firms cannot always capture the full value of new innovations, for they can frequently be copied or enhanced by competitors. These spillover effects may cause insufficient investment in innovation.

Nevertheless, it seems obvious that the free market provides significant, continuous incentives to innovate in order to improve our material well-being. This system has produced enormous amounts of creative innovation aimed at improving our creature comforts. The free market offers the possibility that somebody may take a large risk for a potentially large future benefit because any one of a large number of decentralized decision-makers may choose to innovate. Hence, the decentralized nature of free market decision-making makes it rather dynamic.

Societies vary, however, in how much encouragement they offer innovators. Indeed, firms within a society vary as to whether or not they encourage innovation and what types of innovation they might encourage.

Comparative study of institutions leads to the concept of adaptive efficiency, which grows out of the need to have a concept that works well in a setting where uncertainty prevents accurate calculation of costs and benefits. Adaptive efficiency, as mentioned in this book's introduction, aims at maximizing the number of experiments and providing feedback mechanisms to detect failures after carrying out experiments.

To appreciate the value of the adaptive efficiency model, it might help to think about the value of failed experiments in radical qualitative innovations. First of all, the uncertainties that bedevil all R & D can be especially acute if a radical innovation is involved. A radical innovation may involve a production method sufficiently different from what came before that no one knows if it will work.[12] Or it might involve offering a completely new product or service, so that a producer cannot be sure of consumer interest and acceptance. Yet, radical innovations may offer especially great potential to improve our lives in various ways.

High uncertainty means a high probability of error—in other words, lots of failed experiments. Since we often cannot know ahead of time whether an experiment will succeed or fail, failed experiments are the necessary price to pay in order to have successful experiments. A society that does not experiment, does not adapt, grow, or change. A society that encourages failures, by encouraging experimentation aimed at unpredictable success, will grow, adapt, and change, since some of the experiments will prove successful.

Failed experiments, however, are not just necessary bad things that we have to put up with to have successful ones, they have intrinsic value. Producers can learn from failed experiments, as the Dow example shows. Figuring out what does not work may be an important step toward figuring out what does.

The cumulative results of several inefficient projects combined with some successful ones probably offer great benefits to society. An a priori cost-benefit test that screened out unsuccessful projects in the face of uncertainty would screen out successful ones as well. And decisions based on such a test would eliminate important learning from failures: learning that is crucial to growth and change.

I do not mean to suggest that some screening never has any value. But I do mean to suggest that the value of efficiency does not coincide neatly with the free-market values of freedom, creativity, and growth. Indeed, the tension between pursuit of adaptive efficiency and avoidance of inefficient losses is serious.

The free market offers a model of good (but not great) adaptive efficiency. It does seem to encourage a number of experiments, and it does provide feedback as to how well the innovations perform. Innovations may be financed for some time in hopes of future payoff. But if the public repeatedly rejects an innovation, the company making the new product will cease to do so and the innovating company (or another company) will offer something that better meets the paying public's needs.

The Impact of Material Innovation upon the Environment over Time

The systematic incentive that free markets create to improve our material well-being has serious implications for the environment. It means that anyone who can think of a way to extract resources now serving the needs of plants and animals to make them into products serving human needs has an economic incentive to do so.[13] This systematic incentive to destroy plants and animals harms the ecosystem.

Once an activity harming the environment through resource extraction has become part of our lives, many types of innovations increasing efficiency tend to accelerate environmental destruction. Any technological improvement that increases the rate or scope of extraction methods accelerates the depletion of natural resources. For example, the invention of the chain saw, which greatly increased the efficiency of logging operations, accelerated the destruction of forests. The invention of drift nets has decimated many fisheries by allowing fishermen to catch more fish with less time and effort than ever before.[14]

This means that many increases in market efficiency exacerbate environmental problems. Drift nets and chain saws offer examples of efficiency enhancing innovation, in the sense that they facilitate faster and cheaper extraction of resources and therefore probably lower the cost of related products (e.g., lumber and fish).

This environmental destruction often harms human beings. The ecosystem directly meets vital human needs. For example, wetlands, which provide habitat for large numbers of species, also provide flood control and water filtration services for human beings. When developers drain wetlands to build roads and buildings, they convert land serving the needs of species living in the wetland to human use and destroy the wetlands' flood prevention and water filtration capacities. Researchers estimate the value of services that the environment provides human beings at $36 trillion per year, an amount that is close to the annual gross world product.[15] Destruction of this environment can produce an enormous decline in welfare.

The systematic incentive to innovate in ways harmful to the environment harms human beings directly, as the example of lost flood control and water filtration through wetlands destruction illustrates. Ecosystem destruction also harms us insofar as we care about our fellow creatures or the lands they occupy.

Economic incentives exist to carry out innovations that appear to have some hope of proving profitable, even if doing so involves use of a wasteful process that pollutes the environment. The automobile example illustrates this, for automobiles waste 80 percent of the energy from the gasoline they consume.[16] The possibility of pollution arising from a new process matters little in the free market, unless the legal system introduces regulations or liability that make it matter.

Free Market Incentives for Environmental Protection

The free market provides no systematic continuous incentive to deploy many innovations that improve the environment. An example will illustrate this and facilitate explanation. Solar power plants can generate electricity with no emissions. Assume that a solar plant offers air quality benefits vastly exceeding in value the cost of replacing an existing coal-fired plant with a solar power plant. As long as the cost of generating the solar power exceeds the cost of running the coal-fired power plant, utilities probably will not build the solar plant in a deregulated environment. The utility would derive no direct economic benefit from the improved air quality, even though cleaner air might prevent a lot of deaths and illnesses.

The solar power plant exemplifies an innovation to improve our environmental well-being. But one cannot bottle and sell improved air quality. So the free market provides little incentive to build solar power plants, even if they provide excellent net benefits.

A utility might well choose to build a solar power plant once the price falls below that of building a coal-fired plant and a new plant is needed. But that only illustrates the lack of a systematic incentive to improve the environment. This example of cheap solar power illustrates that the incentive exists to use the cheapest possible method of producing electricity. If the cheapest production method happens to coincide with environmental protection, that incentive will aid environmental protection. If the cheapest method of producing power does not protect the environment, however, the incentive will not protect the environment. Clearly, no systematic incentive to innovate for the sake of protecting the environment exists.

This lack of incentives for environmental innovations will not surprise economists, because it is an obvious corollary of a problem economists widely recognize—the existence of public goods. Consumers do not pay for improvements in environmental quality, because such improvements benefit the public at large, not individuals. Because of this, no individual will pay the full cost of a marginal improvement in air quality, since she shares the benefits with so many others who will not pay. This means that no incentive exists to make environmental improvements for their own sake. It follows, logically, that the free market provides little systematic incentive to innovate in order to find better ways of protecting the environment.

If we look at environmental quality as a good that producers could create by adopting environmental innovations, we can more precisely and more generally identify the dynamic market failure. Producers usually have no incentive to make an efficient innovation improving environmental quality, that is, an investment that yields environmental returns exceeding the cost of making the improvement. Indeed, even an innovation generating clear environmental returns greatly exceeding the associated production cost finds no home in a free market.

The free market, however, does provide a mild pro-environmental incentive for a limited class of investments. Since inputs of raw material have a cost, the free market creates some incentives not to waste raw material inputs. A polluter may increase profit margins by making the same product or process with less inputs of material resources. Notice, however, that this incentive does not obtain when waste avoidance has any net positive cost. If avoiding waste has any cost at all, the free market provides no incentive to conserve raw materials. This remains true even if the cost is very low relative to environmental benefits. Because avoiding waste often does require more precise production techniques, it often does carry a cost. This may explain why we rely, for example, upon power generation techniques that produce enormous pollution loads, in part, because these techniques waste more energy than they produce.

Nevertheless, there will be cases when producers can produce an absolute reduction in waste or raw materials usage and save money thereby. This countervailing incentive may help explain why pollution levels in Eastern Europe and the former Soviet Union are even higher than they are in western democracies. These countries lacked free market economies for a long time and they did not have to consider whether they could make or increase profits by reducing material inputs. The lack of democratic accountability may also have played a large role, allowing communist countries to ignore public opinion favoring environmental protection. It may also explain the willingness of many firms, especially the more innovative ones, to devote resources to pollution prevention.[17] Pollution prevention in a manufacturing process may save money.

Nevertheless, several factors limit free market conservation incentives. First, if commodity prices are cheap, incentives to conserve them may be mild or nonexistent. This means that improvements in extraction technology can lessen incentives to conserve resources in manufacturing. Second, opportunity cost factors may limit the extent to which firms respond to conservation incentives. Firms often will not conserve resources, even resources whose use causes major environmental problems, if they can realize a greater return with other uses of managerial time and capital resources.

Some economists have argued that this countervailing incentive to avoid high materials costs powerfully protects the environment. They claim that as natural resources become scarce, prices will rise, and the price signal will favor conservation.[18]

This claim, however, neglects major problems. First, long before a particular resource is scarce in an absolute sense, its harvesting may cause major collateral damage. For example, a timber company can log an entire forest without making wood a scarce commodity worldwide, as long as other forests exist. Clear-cutting a forest may destroy that particular forest's eco-system, eliminating many species of plants and animals.

Second, plant and animal populations sometimes crash suddenly and unexpectedly. Some fish populations have disappeared because of over-fishing and pollution, before rising prices did anything to save them. Indeed, this has happened often enough that the entire hypothesis of scarcity prices adequately protecting natural resources on a regular basis seems like an economist's fantasy.[19]

Third, in the case of some goods regarded as precious and unique for one reason or another, high prices may hinder conservation. For example, rhino horns, thought to have medicinal properties in Asia, have sold quite well, notwithstanding high prices.[20] Indeed, the rising prices encouraged poaching. Rising prices, however, might have some effect when cheap substitutes are available.

Market inefficiency also prevents realization of money saving conservation opportunities, even when the cost savings are quite substantial.[21] Consider energy efficiency. In many communities, electricity prices are high enough to make a range of energy conservation measures worthwhile economically in an abstract sense. Adding insulation or installing better windows, for example, may have substantial up-front costs, but reductions in the cost of heating and air conditioning mean that these investments pay for themselves in a few years and yield annual savings thereafter. Few landlords who are able to force their tenants to pay for utilities will consider these investments worthwhile. And many tenants have no right to make these improvements and do not expect to live in the same house long enough to realize the savings even if they did. Since we live in an economy with imperfect information, some homeowners that would benefit from this energy conservation do not know about the benefits. If they are not freezing, they do not add insulation or replace the windows. And many homeowners simply lack the capital to make economic improvements, if the payback is not immediate. Some homeowners may lack access to sufficient credit on reasonable terms as well. In short, the real market, not the one the economists model, is very inefficient, because it does not really consider societal efficiency over a long period of time very well. It's much better at realizing short-term cost savings for discrete individuals and firms. And often the individual in a position to make an investment is not the person who might realize the cost savings justifying the investment economically. This is true even when environmental improvements will yield a profit in dollar terms, disregarding completely any environmental benefit.

Even if a substantial economic incentive has existed to destroy the environment in the past, one might ask whether such an incentive will continue to exist in the future. After all, we live, say some, in the post-industrial dematerialized world of the Internet and knowledge-based industries.

Once one gets past the slogans, however, doubts emerge about whether incentives to destroy the environment will disappear. Because of population growth and increased consumption, resource consumption in developed countries increased between 1970 and 1996. Some dematerialization has taken place, if one defines dematerialization as a reduction in material flows per unit of gross domestic product (GDP). The service sector's economy's share of employment and value added to the GDP has grown in the United States, while manufacturing's share of overall employment and value added to GDP has shrunk. But total manufacturing output has *increased* between 1970 and 1996. In other words, manufacturing has expanded, but not as rapidly as the service sector. Because of increased consumption, developed country economies have experienced no dematerialization, defined as an absolute decline in material use per capita.[22] Material consumption per capita has actually increased in some developed countries. And overall material flows will not decline with increasing population, unless material consumption per capita actually falls at a rate greater than population growth.

At this point, no one can authoritatively predict precisely how the Internet, for example, will change the economy and our environment. It would be premature to assume that the Internet will countermand prevailing incentives to destroy the environment.

Amazon.com sells products over the Internet. But physical goods still arrive at customers' doors. It builds large warehouses to contain the goods. It ships the products on trucks from the warehouse to its customers' doors. So cyberspace and the physical world interact and support each other.

Already some products are becoming dematerialized and may become more so. For example, consumers can download books and music from the Internet. This means no shipping, and may imply an environmental improvement for goods that consist of information. Many goods, such as musical instruments, cars, medical supplies, and groceries, however, cannot be transported over the Internet, even if they are sold there. So, this dematerialization may not affect most goods.

If the Internet leads to greater allocative efficiency, however, it might increase competition and lower prices. This might increase, not decrease, the supply of goods that involve environmental harms. Perhaps the Internet will lead to some efficiencies in transportation that will ameliorate environmental impacts. On the other hand, more widespread use of computers implies an increase in energy use and the pollution pumped out when more electricity is generated. The Internet may also spur rapid turnover in devices, like the personal computer, that contain toxic materials and must be discarded.[23] In short, it's a little too soon to conclude that current trends will defeat the general tendency of economic incentives for material improvement to favor destruction of the environment.

I do not mean to suggest that the incentive to innovate in ways that harm the environment constitutes the sole or even necessarily the greatest threat to environmental quality over the long term. Diffusion of existing technology and population growth matters a great deal. Free markets, however, provide a systematic incentive to innovate to improve our material well-being. This incentive often, although not always, encourages environmental destruction.[24]

Government Created Incentives for Environmental Improvement

An episodic incentive for environmental protection comes from environmental regulation. When a government adopts an environmental regulation backed by a civil penalty, regulated companies acquire an economic incentive to make the discrete environmental improvement the government demands.[25] But this incentive, as we have seen, is neither systematic nor continuous. It is not systematic, because the demand for the environmental improvement comes from governments that usually behave in fairly unpredictable ways. It is not continuous, because once a company complies with a government regulation, little incentive exists for further improvement.

Regulation has provided incentives to innovate when it is sufficiently stringent and ambitious.[26] But the incentive to innovate comes episodically, not continuously. During the period that the regulation is in place, companies have an incentive to innovate in order to lower the costs of meeting the regulatory requirement.

But firms have a disincentive to develop and disclose promising innovations that might lead to further improvements in environmental quality, because information about environmentally superior innovations might facilitate the setting of higher standards.[27] And such standards might add to their cost, even if they produce enormous environmental benefits.

This disincentive against pollution control does not necessarily apply to manufacturers of pollution control equipment or to competitors with highly regulated existing industries that have lower polluting approaches to meeting a given need. But these actors face not only the usual problems in securing financing for innovation, but the additional problem of government regulation serving as the primary determinant of demand. Uncertainty about future regulation can discourage financing of environmental innovation, because government failure to tighten standards sufficiently to create a market for innovations can shut down the market.[28] Indeed, during the recent Internet boom, when venture capital for material technological innovation in the United States was relatively abundant,[29] venture capital for environmental technologies dropped to minuscule proportions, notwithstanding the existence of regulatory programs increasingly reliant upon emissions trading.[30]

Especially little incentive exists for basic environmental innovation that is uneconomic in the short term. Recall that the free market often encourages material innovations that lose money in the short term in hopes of future gains. An entrepreneur or a company might underwrite the expense of an uneconomic innovation to meet material needs in the belief that the costs will fall as experience is gained or that the innovation will prove transformative enough to gain a large market. Firms, however, rarely pursue an expensive environmental improvement for the sake of future profits.

Government finds it very difficult to write regulations stringent enough to create markets for technologies that either do not work properly yet or are expensive in the short term, even if they promise enormous long-term benefits or falling costs in the future. Government officials may lack the self-confidence to make judgments about what industry can accomplish in the future with unproven techniques. They may, often rightly, regard themselves as lacking sufficient expertise to make such judgments. Even a visionary and knowledgeable government official would face fairly daunting pressures that discourage risk taking of the kind that occurs when the private sector pursues a difficult material innovation. Courts generally require a government agency to explain why it believes that a technology forcing regulation will succeed.[31] While the courts have shown some understanding that any attempt to force technological development must rest upon a willingness to move forward with less than perfect information, they have also reversed technology forcing decision if they found the explanation inadequate.[32] The agency may face heavy political and legal pressures to avoid drastic changes that may harm the interests of companies or individuals benefitting from existing approaches.[33]

Hence, the free market economy provides continuous incentives for material innovations that may harm the environment. Our society provides, at best, only sporadic and uncertain incentives for environmental innovation. It often provides no incentive at all for risky innovation aimed at radical environmental qualitative innovation, which would be very important to our future. A dynamic exists that may tend to favor environmental destruction over the long term, unless government has sufficient capacity to compensate for this dynamic. The next chapter evaluates the capacity of government to make compensating decisions by comparing government and private decision-making structures.

7

Decision-Making Structures

Incentives to innovate do not actually change how we do things until people respond to the incentives. To begin to analyze responses to incentives to innovate, it helps to imagine two entrepreneurs—one with an innovation improving material well-being in some fashion, and another with an innovation improving environmental quality. We can then ask who makes the decision about whether to employ the innovation. Once we identify relevant decision-makers we can describe the processes that govern decisions about deploying these innovations and gain some understanding of how relevant individuals and institutions will respond to the economic incentives described in the previous chapter.

The people and institutions who make decisions about deploying innovation will make them within the constraints of bounded rationality.[1] We can compare the institutional actors who make key decisions about material and environmental innovation. In particular, we can observe the habits, routines, procedures, and orientation that create the specific bounds upon their rationality.[2] This leads to some predictions about the propensity of relevant decision-makers to deploy innovation to meet material and environmental needs.

Decision-Making to Improve Material Well-Being

An individual with a new product or a new process may make an individual decision about whether to employ a material innovation. The innovator must simply decide whether the possible rewards justify the hard work and financial sacrifice that might be necessary to bring an innovation to market. If she has invented a product or service, she may try selling that product or service directly to customers.

Implementation of many innovations, however, requires large capital investments. If the innovator needs financial backing, a more complicated decision-making structure applies.

An innovator may need to persuade investors that they can make money by financing implementation of her idea. Jeff Bezos, for example, persuaded venture capitalists to finance Amazon.com early on and then persuaded individual investors to buy stock in the company.

The people who will make decisions about whether somewhat capital-intensive innovations are realized stand to gain, not lose, from the success of the innovation. They lose out only if the innovation fails. Hence, they, like the innovator herself, have a substantial incentive to back promising ideas, even though success is not assured.

Technological change, however, often creates losers as well as winners. For example, if Amazon.com succeeds, competing booksellers may lose out. The potential losers from successful technological innovation, however, should have little direct say in whether an innovative project receives financing. Thus, competitors' fears about losing business should not directly influence potential investors' decision-making processes.

To be sure, potential investors may consider competition's effect on a potential innovation's prospects as they decide whether to invest in a new firm or technology. But they normally will not treat the competitors' potential losses from the innovation's success as a negative factor in the decision-making process. Rather, absent pressures from competitors, the potential investors will only be concerned with the competitors' ability to make the innovation (or the innovator) fail in the marketplace. If a particular investor responds favorably to a competitor's pressure to avoid financing a new venture introducing a promising innovation, competing investors may finance the new venture. Thus, the decentralized nature of market processes tends to limit the influence of existing industries on decisions to finance new innovations.

Decisions that require investment may necessitate some formal analysis. For example, venture capitalists often like to see a business plan that sets out how the entrepreneur proposes to transform an innovative idea into a viable business.[3] While these documents can be fairly long, most contain an executive summary of no more than three pages, which potential financial backers often use as a screening device.[4]

An innovator might also seek to persuade an existing firm to use or market her invention. An innovation that lowers the cost of producing an existing good or service finds its market among existing producers, not among consumers. And an innovator with a new product or service may want to sell it to a firm that has the capital resources to mass-produce it and market it. Existing firms may purchase an entrepreneur's invention, but they also may actively develop innovations in-house or copy a competitor's innovation (copying, however, might be better thought of as diffusion of technology, rather than innovation).

Firms generally will have a more bureaucratic process to decide about what innovations to employ.[5] This may require formal analysis and review by management.

The firms or individuals considering an innovation, however, control the decision-making process. A venture capital firm can decide how much time to devote to any particular proposal and may demand fairly short business plans. A firm may likewise ask for the information it considers most important and limit the length of written analysis. It may limit the length or frequency of meetings in order to conduct enough thoughtful discussion, without wasting time on trivial considerations. In short, those deciding whether to employ a new innovation may structure a decision-making process that provides for adequate consideration of the most important information without dwelling on trivia or addressing the same issues over and over again.

The decision-makers will consider both the potential benefits of the innovation and its potential costs. If the benefits appear great, many decision-makers will be willing to take substantial risks to bring an innovation to market. This implies that uncertainty about costs or current high costs does not necessarily defeat a project that promises sufficient benefits to consumers to generate substantial revenues.[6]

Furthermore, the large number of potential decision-makers in a competitive market can increase the chances that someone will take a large risk in order to realize large future benefits. Thus, if one or two companies behave in a risk-averse fashion, an entrepreneur may still take the risk to bring an innovation to market.[7] This, in turn, limits the conservatism of companies that might like to forego innovation in favor of the status quo, because they may lose out if someone brings a good innovation to market.[8]

This decentralization offers the promise of a large amount of innovation over time. Since many different actors can deploy innovations, many innovations over time might find a market.

On the other hand, in less competitive markets dominated by mature firms that can create substantial barriers to entry, innovation may suffer. In such cases, firms may find it in their interest to keep doing what they do well and avoid radical change. But even here, barriers may fall and companies must consider that.

Similarly, individual firms may only consider innovations that fit their existing business in some way, that take advantage of existing skills at the firm, that resemble products or services that the firm already offers, or that fit with some strategic plan to meet the needs of existing or targeted customers.[9] These rules of thumb do not substantially constrain innovation overall, because the large number of firms make substantial opportunities available for a wide variety of material innovation, but they may constrain particular types of innovation (for example, alternatives to very well established capital intensive technologies).

The free market offers decentralized flexible decision-making by people who will benefit from an innovation's success. This provides a reasonably hospitable environment for material innovation.

Decision-Making to Improve the Environment

An entrepreneur with an environmental innovation confronts a vastly different decision-making process. Her invention, if it is a consumer good, does not offer any advantage that any one consumer can capture. Hence, she probably will find it difficult to just decide independently to produce the good and make money that way, especially if the good costs more to produce than competitive products (if it costs less, then it is not just an environmental innovation, but also a material innovation, and regular free market incentives apply). Some producers have tried to market the environmental attributes of their product as a means of persuading consumers to buy their products. But this sort of green marketing may require capital resources that an individual does not possess and probably offers a lesser opportunity than an invention that lowers costs or offers a material improvement to an individual.

A prominent apparent exception to the rule that individual environmental innovation is quite limited—the production of organic produce—turns out, upon analysis, not to be an exception. Organic produce commands a premium price primarily because it promises to protect the health of the purchaser in proportion to how much she purchases (or more precisely, how much she substitutes organic produce for produce bearing pesticide residues). While the environmental benefits—the protection of land and water from the effects of pesticide use—may enhance the product's appeal, organic produce actually offers a direct material benefit to each individual purchaser, not just a benefit to the environment (or people) in general.

Many environmental innovations of consequence, however, will require capital to produce, and many involve qualitative changes in production processes. Such changes may produce no particular benefit to the individual consumer of the product as a consumer, but rather provide benefits to the society as a whole (or at least all who come into contact with the relevant pollutants). An entrepreneur must then persuade an existing producer or backer to finance the innovation.

Imagine an entrepreneur who wishes to sell more expensive cars that pollute less seeking backing from a venture capital firm. Her business plan (if it's honest) would have little appeal to venture capitalists, because she promises to deliver existing consumer benefits at a higher cost. She probably cannot persuade a capitalist that the environmental benefits of low emission vehicles will generate higher sales, so she may have difficulty obtaining funding. If she tries to sell an innovation improving environmental quality to a firm (for example, a qualitative environmental innovation in production processes), her prospects are even worse. As explained previously, in many cases innovations reducing environmental impacts will reduce rather than increase the profits of the enterprises employing them. So, the free market mechanisms governing innovations improving our material well-being do not lead to the adoption of many desirable environmental innovations.

In particular, private firms may not consider the benefits of environmental improvements especially relevant to their decision-making. These benefits produce no revenue for them, so they may not matter. This means that the decision-making process for deploying material innovation generally does not apply to environmental innovations. The possibility of incurring any cost to realize environmental improvements, no matter how valuable the improvements and how minuscule the costs, defeats most potential environmental innovation in the free market.

The private decision-making process for material innovations might apply to those environmental innovations that generate incidental cost savings (for example, more energy efficiency in a production process). But most innovations of any kind generate added cost. Most qualitative material innovations generate revenues exceeding cost eventually, but the producer often must incur costs, not enjoy savings, in order to deploy the innovation. While the free market provides sufficient incentives for many economically justified material innovations, it provides sufficient incentives for only a tiny fraction of the many economically justified environmental innovations.

If an innovation offers an environmental improvement as its sole or major attraction, then demand for that innovation will usually depend upon government action. If the government demands a sufficient environmental improvement to create demand for the innovation, then firms may decide to use the innovation. If the government makes demands too weak to make the innovation necessary or demands nothing, then the innovation finds no market.[10]

This implies that governmental decision-making structures influence choices about whether to implement an environmental innovation.[11] The government, in effect, indirectly decides whether or not to employ environmental innovations through its regulatory decisions. Decisions to regulate strictly enough to require or at least invite innovation indirectly create financing for innovations. Firms facing regulatory requirements must spend money on compliance and may have to pay for implementation and sometimes development of innovations. If the standards require more than their existing processes can deliver, they must innovate. If the alternative to innovation involves use of an expensive existing technique to comply with strict standards, then a substantial incentive to innovate will exist.

The key role of government in generating environmental innovation means that comparison of private sector decision-making structures about material innovations to public decision-making structures about environmental innovations aids understanding of responses to incentives to innovation. This comparison remains valuable, even though thoughtful political scientists have pointed out that the public/private distinction blurs at the margins.[12] We shall see that the decision-making processes of public and private organizations relevant to environmental innovation do vary.

Congress sometimes writes specific standards limiting emissions in important polluting sectors. Because of the sheer number of pollution sources, Congress more frequently delegates most specific standard setting decisions to EPA and its state counterparts. This means that Congress and administrative agencies have key roles in deciding, at least indirectly, whether to deploy an environmental innovation.[13]

Neither the Congress nor an administrative agency has a direct interest in successful environmental innovation. Unlike the key private sector decision-makers choosing material innovation, these actors will not profit financially from successful environmental innovation.

The environmental entrepreneur becomes a peripheral figure in the formulation of environmental policy. The policy debate is largely a debate between advocates of environmental protection and existing polluters (often well established industries seeking to minimize any interference with their institutional routines). This means, in effect, that someone with an environmental innovation confronts a decision-making process dominated by rivals. So, for example, makers of fuel cells will find themselves vying with coal producers to influence decisions that might lead to deployment of their innovation. This contrasts with the decision-making process for material innovation, where the innovator can often expect to exclude consideration of competitors' welfare from the decision-making process.

This influence of existing industry may seem surprising, since Congress often legislates with innovation in mind. For example, Congress passed the Clean Air Act in the face of industry claims that pollution reduction was impossible, because it believed that industry would respond to strong mandates by innovating to reduce pollution. While the Congress that passed the Clean Air Act contemplated that industry would either meet health-based standards or shut down, implementation has relied primarily upon regulations inducing existing industry to adopt incremental innovation or readily available technology and keep operating. Environmental statutes have rarely forced major industries to shut down in favor of new industries offering radical environmental innovations.[14]

Although many statutes allow the agency to force technology and demand innovation, an economic dynamic tends to give existing industries, often heavy polluters, a major role in the implementation of the environmental statutes. Companies that have converted resources to human use receive rents from those who want or need their products. Any person who drives a car and heats her home, for example, sends cash to utilities and oil producers and refiners on a regular basis. These companies use a portion of this revenue to hire lawyers, lobbyists, and scientists to argue against strict regulation. Even people who want stricter environmental regulation finance sophisticated technocratic efforts to weaken environmental regulation, on a massive scale, through their bill payments. This economic dynamic influences both Congressional and administrative decision-making.

This point should not surprise anybody familiar with the public choice literature, which predicts that well organized groups will have disproportionate influence upon political decisions. But the economic dynamic analysis clarifies several important points. First, the amount of financial resources at the disposal of organized groups might matter. Second, over time, innovators finding new ways to convert natural resources into products for human consumption will acquire the resources needed to influence government decision-making.

Some of the legal and political science literature claims that politicians have incentives to take credit for ambitious statutory goals, while writing ambiguous provisions for implementation that provide opportunities to perform casework on behalf of constituents that might undermine achievement of these goals.[15] Thus, for example, the Clean Water Act has a zero discharge goal, but the details of the statute tend to require only "best available technology."[16] This leaves open an opportunity to debate what precisely the best technology is. And almost all of the implementation provisions allow for (or even require) consideration of cost, thus providing further opportunity for debate. Congressional representatives and presidential staff can curry favor with powerful industries by intervening in rulemaking proceedings on their behalf.[17] While one can debate whether the ambiguities and moderating provisions found in the statutes represent strategic behavior by legislators or an inevitable feature of law-making governing complex issues, surely casework does go on. The economic dynamic I have described predicts that the casework will more often serve well-established existing industries than fledgling environmental entrepreneurs.

Agencies have substantial incentives to avoid decisions that will disrupt existing industries, for such disruption can lead to political pressure upon the agency. For this reason, an agency may prove reluctant to demand innovation, especially radical environmental innovation that may change the winners and losers in the marketplace. Unlike at least some private companies in a position to deploy innovation, the agency may care much more about continuity than the benefits of change.

The biggest existing polluters receive most of the regulatory attention. Agencies do not regulate by figuring out what clean technologies are out there and then asking the developers of these technologies what regulations would help them find a market, hopefully at the expense of dirtier competitors. Rather, regulators focus on the dirtiest existing industries and use regulatory proceedings to get them to clean up. So, a major party in most rulemaking proceedings has a great financial stake in continuing the use of current technology.

Many environmental entrepreneurs will not participate at all in rulemaking proceedings like this. Many of them will get their innovations to market, if at all, only through selling their innovations to the regulated companies. They are therefore frequently anxious not to alienate their potential customers by advocating more stringent limits on pollution based on the capacity of their innovations.

Many innovators also lack the financial muscle of a well established industry. Hence, they often lack the resources to lobby vigorously for strict regulation, even if they are willing to participate.

Regulated companies, usually entrenched opponents of environmental innovation, participate very actively in rulemaking proceedings. These rulemaking proceedings follow a notice and comment structure rather than a private business plan model. Instead of the innovator contacting the decision-maker, the decision-making agency issues a notice to the general public that it will conduct rulemaking and invites comments upon a proposed rule.[18] Regulated industry files thousands of pages of comments in typical rulemaking proceedings. The volume of comments from regulated parties regularly far outstrips that of any other participating groups.[19] The regulatory agencies have a legal duty to respond adequately to any

significant comment. Since industry has the resources to make voluminous legally and scientifically complex arguments, responding adequately
requires an enormous investment of time and resources.

While the private decision-makers' decision to adopt an innovation is
usually final, agency decisions to demand improvements that might lead to
innovation become the subject of judicial review. Any failure to respond
adequately to any single argument deemed significant by a reviewing court
can result in judicial reversal of an agency decision. And agencies have no
real way of knowing which comments a court will focus on as significant.[20]
Regulated industries bring the overwhelming majority of suits seeking to
reverse agency rulemaking, so agencies face substantial pressures to carefully consider their views.[21] Again, this greater volume of input is a product of the economic dynamics of realizing revenue from polluting
processes.

Congress created deferential standards of review in order to avoid too
much second-guessing of agency decision-making.[22] Thus, many statutes
only allow reversal if agency decisions are contrary to law, arbitrary, capricious, or an abuse of discretion. But the courts have frequently construed
these standards to require a "hard look" at agency decision-making. In taking a hard look courts sometimes find it difficult to distinguish between an
arbitrary decision (which the court should reverse) and one that strikes
them as poorly reasoned (which they should uphold if non-arbitrary). Since
courts are expert critics of reasoning and have very little understanding of
how the complexity of the information the agencies must consider limits
agencies' abilities to justify decisions simply and plainly, some courts find
grounds for reversing agency decisions frequently.[23] In any case, the risk of
reversal is ever-present and makes agencies very reluctant to make decisions, especially demanding ones that will excite intense opposition.[24]

The Supreme Court has held that courts should defer to reasonable
agency views about the meaning of ambiguous statutes.[25] But this instruc-
tion has not greatly constrained reviewing courts.[26] Furthermore, the
Court's most recent pronouncements in this area have cast doubt about the
scope of this deference.[27]

In spite of this economic dynamic, the government occasionally makes
decisions that demand innovation. Some countervailing pressures apply to
the agencies. Some environmental groups have the capacity to generate

technically sophisticated comments, litigate agency decisions, and meet with government officials. Since the public generally supports stringent environmental protection,[28] public pressure may sometimes persuade regulators to adopt strict standards. Furthermore, the statutes governing the agencies sometimes specify stringent regulatory requirements at a fairly high level of detail. For example, Congress has often set numerical limits applicable to the automobile industry to force the development and application of pollution controls. Although government sometimes demands sufficient environmental performance from industries to encourage incremental environmental innovation, it very rarely demands radical technological change displacing an existing industry.

Administrative agencies lack the degree of control over their own decision-making processes that private decision-makers considering material innovations enjoy.[29] When an agency writes regulations, Congress generally requires the use of notice and comment rulemaking, agency response to all significant comments, various mandatory forms of analysis, and the availability of judicial review. All of these requirements aim to serve admirable purposes, such as the facilitation of open public participation, promotion of reasoned decision-making, and compliance with statutory mandates. To be sure, agencies often (although not always) enjoy some flexibility in deciding whether to adopt rulemaking or proceed case by case in setting the length of the comment period, in determining whether to hold hearings, and in deciding who within the agency will lead rulemaking.[30] Nevertheless, Congress and the courts have imposed a significant set of procedural requirements upon agencies that substantially limits their flexibility in many contexts.

By contrast, private decision-makers considering material innovations enjoy the flexibility to structure their own decision-making processes. They face no judicial review, may limit participation to those whom they consider most helpful to them, conduct whatever analysis they consider appropriate (or none at all), and respond to participants only if they wish to.

This entire comparison of government to private sector decision-making may appear inappropriate, however, for an important and interesting reason. The private actors deciding whether to deploy material innovations are producers. They decide what to supply. Government decision-making that may induce environmental innovation creates market demand for innovation. The government role in environmental innovation resembles that of the consumer in the world of material innovation, not that of the producer.

This asymmetry should not matter, because the key initial actors in deciding about innovation have been properly identified as the producers in the material context and the government in the environmental context. This asymmetric market structure, however, yields an important insight. Supply drives material innovation, whereas demand drives environmental innovation. While, ultimately, the success or failure of an innovation in both contexts involves a matching of supply and demand, the introduction of material innovation begins with a producer deciding that the supply, often coupled with advertising, will induce or meet demand.[31] Environmental innovation generally begins the other way around. Demand from the government may induce demand for innovation from regulated firms, which will generate a supply of innovation from other firms or from within the regulated firm itself.

The pressure from both sides combined tends to paralyze administrative agencies. Professor Thomas McGarity of the University of Texas School of Law has described this problem as an "ossification" of agency rulemaking, leading to a very slovenly pace of regulation under statutes containing mandatory deadlines (which are almost never met) and almost no regulation when no statutory deadlines apply.[32] This ossification claim has a wide following among administrative law scholars.[33] The possibility of reversal on points that appeared trivial at the time of rulemaking tends to paralyze agency decision-making. The most recent critique of the ossification claim argues that the agencies frequently succeed in passing implementing regulations fairly soon after judicial remands.[34] The claim ignores the fact that some judicial remands have led to some very striking delays and regulatory failures in regulating very serious health hazards, including the failure to ban asbestos[35] and long delays in regulating benzene,[36] although both present serious health hazards. Furthermore, if correct it would only lead to a recognition of some limits to the ossification point, not to its refutation. Ossification would have to be recognized as a slovenly pace of regulation, rather than a permanent paralysis.

In spite of the well-recognized inefficiency of agency rulemaking proceedings,[37] most recent regulatory reform recommendations have sought to make the process even more inefficient by adding further analytical requirements.[38] The combination of judicial demands for a more complete record for review and first executive and later congressional demands for cost-benefit analysis has increased the amount of paperwork agencies must complete as part of the rulemaking process. The so-called "informal rulemaking process" has become much more rigid and formal than the process potential stimulators of innovation in the free market would choose, and is, therefore, terribly slow.

Since innovators can go out of business as they wait for a market to develop, the uncertainty and lethargy of the regulatory process matters to innovation. Moreover, an environmental innovator seeking a national market, unlike a material innovator, does not have a wide variety of avenues open to create the opportunity to deploy an environmental innovation. The states can at times offer an alternative venue. But their rulemaking procedures often follow the federal model.

Both the states and federal government also have limited regulatory capacity. They generally have a wide environmental agenda, limited budgets, and a time-consuming process that makes nearly every move a long, drawn out battle. This means that government, at least government acting as an administrative decision-maker, cannot hope to offer the variety of opportunities to environmental innovators that a decentralized free market offers to material innovators.

I do not claim that regulated industry has wholly captured administrative agencies.[39] Opposition to disruption involves something much more mild than complete capture of an agency, defined as making an agency entirely subservient to industry's will. Political scientists generally recognize that the regulatory reform movement has slowed and compromised environmental regulation and sometimes empowered regulated industry, but they have not clearly demonstrated that this amounts to economic capture of the administrative agencies.[40] And some studies have concluded that industry comments in the rulemaking process have only a limited impact upon the final rules.[41] Nevertheless, the process has become cumbersome enough to slow government decision-making and to impede regulation demanding substantial innovation.

Regulators have sought to increase reliance upon negotiated rulemaking to avoid the pressures of adversarial and often litigated rulemaking. These regulatory negotiations, while often successful in satisfying the parties, sometimes fail and usually take a significant amount of participant and agency time.[42] Moreover, a negotiated solution rarely rewards environmental entrepreneurs ready to supplant the basic technology undergirding an industry. Rather, the kind of regulation that wins approval from both environmentalists and industry manages a substantial enough environmental improvement to satisfy environmental groups, while disrupting an existing business too little to disturb the regulated industry.

International decision-making also may affect the market for innovations. Treaties, however, come about because of the unanimous consent of countries. This means that a proposed treaty text must reflect limits to environmental destruction that can win the voluntary consent of all relevant countries. Often, some countries oppose demanding limits on a particular threat to the environment because of fears about the impact to their economies of measures that might threaten an existing industry with high costs or, worse yet, abolition. In such situations, it will often prove easier to adopt a limit that does not stringently restrict environmental destruction. The requirement of unanimous consent may generate pressure to adopt a lowest common denominator position.

Existing industries often impress their views on their countries' government officials. And many industries tend to favor proposals that do not require innovations that would radically change how they do things or, worse, allow a competitor with a newer technology to displace them.

A good example of this comes from the history of the Bush family and the Framework Convention on Climate Change. The overwhelming majority of developed countries favored a firm commitment to reducing greenhouse gases on the eve of negotiation of the Framework Convention. But the United States did not. Companies operating electric utilities, running oil refineries, and manufacturing cars vigorously and successfully lobbied the first Bush administration to oppose reductions in greenhouse gas emissions. The resulting compromise adopted an "aim" of stabilizing emissions over a fairly long period of time. Years later, the Kyoto Protocol to that convention again involved a lowest common denominator position, adopting cuts less stringent than those advocated by many European countries. President Bush's son, shortly after his election, rejected even the fairly modest demands of the Kyoto Protocol.

The international decision-making structure makes agreement on standards demanding substantial innovation unlikely. The Montreal Protocol, however, created a substantial market in substitutes for ozone depleters banned under the protocol.[43] But the Montreal Protocol stands out as an anomaly across the landscape of international environmental decisions. Indeed, a careful look at the competitive dynamics helps explain the anomaly. The United States strongly supported a ban on certain ozone depleters after DuPont Industries announced that it had developed profitable substitutes for the ozone depleting substances it had made in the past.[44] And the Protocol's implementation respected the potential as well as the limits of the U.S.-DuPont position. The administration of technology transfer provisions under the Protocol tended to favor replacement of ozone depleting substances with DuPont and other chemical industry favorites, even though these substitutes also depleted the ozone, albeit at a much slower rate.[45] More radical substitutes favored by some environmental groups won little support in the efforts to spur technological substitutes for the banned substances.[46]

This does not mean that DuPont's announcement alone explains the Montreal Protocol or its implementation. Vigorous lobbying by environmental groups, a solid scientific understanding of ozone depletion, and the Reagan administration's eagerness to redeem a tarnished environmental reputation all played a role. But one cannot be sure that the international community would have banned ozone depleters had the chemical industry remained opposed to it.

Often, existing industry has a stake in existing techniques that leads them to oppose limits forcing innovations upon them or, worse, favoring fledgling competition. These positions can influence at least some countries. In general, international consensus decision-making provides an awkward vehicle for stimulating environmental innovation.

Moreover, international environmental agreements produce no innovation and no environmental improvement directly. Rather, international agreements establish state commitments to protect the environment. No environmental improvement comes until the relevant nations take further actions inducing private parties to carry out measures protecting the environment.[47]

This means that the implementation of international agreements relies upon the same legislative and administrative processes that make implementation of domestic environmental programs so slovenly.[48] As a result, the implementation of international environmental agreements is often (but not always) slow, uncertain, and unsatisfactory.[49]

Both international and domestic environmental decision-making processes give existing industries a substantial role in deciding whether new environmental technologies will be employed. The purveyors of environmental innovation (unless they happen to be an existing regulated industry) have little say. This contrasts rather sharply with the decentralized decision-making structure governing free markets' innovations to improve material well-being. That decision-making structure does not legitimate participation of opponents who could lose if new technology succeeds. The decision-makers stand to make money from successful new technologies' introduction, and the innovator herself has a central role.

Furthermore, while some bureaucratic processes apply to financing decisions within firms and in the capital markets, firms can act more quickly than regulatory agencies. Regulatory agencies face enormous pressures to regulate slowly or not at all.

In short, the process governing deployment of environmental innovation is far more centralized and cumbersome than the process governing deployment of material innovation. The next chapter explains why this matters.

8

The Shape of Environmental Problems: Growth, Decentralization, and Change Over Time

We need a more dynamic system of response to environmental problems, because a powerful dynamic tends to create and magnify environmental problems over time. Increasing population and consumption tend to increase resource consumption and pollution, thus worsening environmental problems. Incentives for innovation increasing environmental destruction may exacerbate this problem, especially the incentive to develop more efficient extraction technologies. Many of the most serious unaddressed problems have numerous decentralized sources,[1] which raises questions about the capacity of a relatively centralized regulatory system. Finally, the regular emergence of new problems challenges the creativity and dynamism of regulation. This chapter will discuss population, increased resource consumption, environmentally destructive innovation (briefly), decentralized problems, and periodic emergence of new environmental problems in turn.

All institutions must adapt to changing conditions. Describing the changes the environmental legal system must face aids evaluation of how suited the current system is to the problems that will beset us.

Factors Increasing Pollution over Time

Increasing Population
Some very fundamental dynamics tend to accelerate environmental destruction over time. World population has increased from 2.5 billion to 6.1 billion just since 1950.[2] While the pace of this growth has slowed, the increase continues, especially in developing countries. Demographers predict that world population will increase by 2.8 billion over the next 50

years, bringing total population to 8.9 billion.[3] Even a slower growth rate applied to the enormous current population base implies a significant numerical population increase. Notice that the projection mentioned above implies that more people will be born in the next 50 years than in all of the time preceding 1950.

Since all of these new people will consume resources and generate pollution, growth in population normally implies a growth in environmental problems. As we shall see, many of these problems are global in nature and affect people in countries that do not themselves experience large population growth. Thus, for example, rates of industrialization in China can have a large role in determining whether climate induced sea level rise inundates Louisiana.

Increasing Per Capita Resource Consumption

People in developed countries consume far more resources than any large group of people in history. People in developing countries generally wish to enjoy similar opulence. Some developing countries that have already increased their wealth have experienced a significant increase in environmental destruction. China and Thailand, for example, have grown enormously economically, so Bangkok and Beijing have some of the most unhealthy air in the world. Many developing countries have also denuded forests, ruined land, and fouled water supplies at a prodigious rate. If developing countries increase their population and consumption over time, as most expect them to, they will use many more natural resources and pollute more in the future than they do now. This will accelerate both local and global environmental harms over time.

Increasing wealth may increase capacity to address environmental problems. Developing countries often lack sophisticated environmental bureaucracies capable of regulating pollution and exploitation of natural resources. As these countries become wealthier, they typically increase their regulatory capacity. The question of whether that capacity can keep up with the pace of environmental change, driven by potentially large increases in consumption and population, remains open. This question of developing country capacity is important not just to the future of developing countries, but to the future of developed countries as well, since some of the pollution developing countries may produce in the future will contribute to global problems.

Some clues about the capabilities and limits of the current economic dynamic come from developed countries. These clues suggest that matters beyond the purview of the regulatory system tend to undermine efforts to cope with environmental change, even in countries with well developed regulatory systems. Saying that they undermine those efforts, I do not mean to suggest that they completely defeat all efforts. In fact, some progress has occurred in spite of this problem. But even in the United States, which has in the past led the development of environmental law, progress has fallen short of the ambitious goals that Congressional representatives have chosen for the U.S. environmental system and sometimes short of alternative less ambitious goals favored by many academics (such as optimal pollution levels for society).[4]

In general, developed countries have ameliorated many of the environmental problems that have come with increased wealth and production, but solved very few of them. Air pollution levels remain unhealthy across much of the United States and Europe.[5] Few bodies of water in developed countries provide water adequate for drinking, swimming, fishing, boating, and recreation. While we have made significant progress in some places, we have never fully recovered from the onslaught of problems that came with industrialization.

Developing countries often have denser populations than developed countries. This implies a greater magnitude of environmental problems as living standards rise and a corresponding greater challenge for their environmental efforts. While it is possible to imagine improved regulatory capacity ameliorating pollution increases in developing countries as they become more industrialized, regulatory systems cannot prevent substantial environmental degradation driven by increasing population and consumption without becoming much more dynamic.

Technological Diffusion and Innovation
Absent dynamic change, technological diffusion can spread environmental problems to more countries, thereby exacerbating pollution. For example, the widespread introduction of automobiles among very dense populations in Asia leads to enormous new pollution problems. Some of the relevant pollution increases, such as increases in greenhouse gas emissions, will have worldwide effects.

Similarly, ongoing incentives for innovation can increase resource consumption and pollution. Indeed, there is a dynamic which systematically encourages innovations increasing resource consumption, with some tendency to encourage increased pollution as well. Chapter 6 described this dynamic in some detail.

Decentralized Pollution Problems

Developed country regulatory systems have partially succeeded in addressing pollution from large, prominent sources. Consequently, most further progress will require regulation of numerous smaller sources, or fundamental technological change for larger sources. The problem of decentralized pollution sources poses an enormous challenge for a relatively centralized regulatory system.

Most of our clearest success stories arise from efforts to regulate problems stemming from relatively few large scale sources. For example, EPA realized an enormous improvement in public health when it phased lead out of gasoline.[6] This action focused on one chemical in one industry with enormous economic capacity. Even with respect to lead, the United States has moved slowly to address the remaining problems in old buildings with lead paint, leaching of lead from old pipes, and other sources. Evidence is mounting that lead remains a serious problem in some urban areas. Nevertheless, we have made significant progress in greatly reducing lead exposure and the associated risks.

Similarly, the international phaseout of ozone depleters has largely succeeded.[7] Only a handful of companies in a few developed countries made the key chemicals. While some issues remain about some ozone depleters that have proven more difficult to phase out (such as the pesticide methyl bromide),[8] and smuggling remains a problem,[9] scientists now predict that the ozone layer will recover from the earlier environmental assault, thanks to the phaseout of several potent ozone depleting substances.[10]

But the regulatory process, even in advanced, developed countries, has great difficulty addressing problems that arise from numerous diffuse pollution sources. Although many sources of pollution are small, small sources can be cumulatively significant but difficult to regulate. Restrictions on the activities of consumers and small businesses tend to encounter strong political resistance. In addition, monitoring pollution from such sources is often not feasible. This means that efforts to limit it must take place with especially great gaps in information about the regulations' likely effects. This can make enacting regulations difficult. And enforcement against many small sources can tax the resources of governments if regulations are enacted. Society cannot effectively address problems stemming from numerous diffuse sources without enormously widespread changes. Even the most advanced countries find it difficult to effectively induce enough pollution sources to fully address serious environmental problems.

Consider tropospheric ozone, the principal contributor to "urban" smog. Ozone seriously damages lungs, contributes to asthma, harms trees and crops, and corrodes buildings, cars, and other structures. It has traditionally been considered an urban problem, but unhealthy levels of ozone affect millions of people across wide areas that include many rural communities. An enormously large class of chemicals called volatile organic compounds (VOCs) combine with a smaller class of compounds called nitrogen oxides (NO_x) and sunlight to form ozone.[11] VOCs come from cars, oil refineries, paint, use of consumer products (e.g. shampoo), furniture manufacturing, chemical manufacturing, body shops, print shops, dry cleaners, boats, planes, trucks, buses, and hundreds of other sources. NO_x comes primarily from fossil fuel burning. A large amount of activity involves fossil fuel burning, including operation of cars, buses, airplanes, boats, locomotives, lawn mowers, weed trimmers, and power plants.

EPA and the states have regulated many of the larger sources of NO_x and VOCs, but have not been able to produce clean air in most major cities after thirty years of effort. Many of the smaller sources remain almost wholly unregulated in many states. Europe likewise has succeeded in regulating some of the larger sources, but has made less progress in addressing smaller sources.

Some areas of the United States have met air quality standards, but generally not the largest cities.[12] And EPA now considers many of the old air quality standards inadequate.[13] Under new standards for urban smog, few areas will have healthy air as defined by EPA. While air quality has improved, EPA has not generally brought pollution levels down to reasonably safe levels.

One of the problems at the root of the failure to bring healthful air quality to all Americans illustrates the importance of factors beyond the effective reach of the regulatory system. Congress has mandated vast reductions in automobile pollution over the years. Today's cars pollute far less than vehicles thirty years ago.[14] But car use has increased by orders of magnitude.[15] Many families now have two cars and these cars travel many more miles than they did three decades ago. EPA regulated the emissions per automobile, but the economy continued to change in ways that favored more vehicle use. The increased vehicle use wiped out some of the progress (although not all of it).[16]

The history of water pollution control in the United States likewise demonstrates the importance of factors beyond the effective reach of the regulatory system. We need water for many purposes. We drink it and we use it for swimming, fishing, and recreation. Many water bodies have become so polluted that we have had to curtail many of these uses. Water pollution has also forced federal, state, and local governments to spend more than $128 billion to treat municipal sewage.[17] EPA has estimated that meeting additional water treatment needs would cost more than $110 billion (in 1990 dollars) through the year 2010.[18]

EPA has addressed these problems fairly vigorously by regulating large factories—point sources in the jargon of the Clean Water Act.[19] But the most important remaining problems stem from numerous diffuse pollution sources that do not even collect their discharge at a single point.[20] These non-point sources include runoff carrying pesticides from farms, manure from animal feeding operations, overflowing sewers, oil and gasoline from roadways, and a host of other sources. These non-point source pollution problems have proven fairly intractable. EPA has had to rely upon the states to address non-point source pollution, and most states have done little about it.[21] Because of this, a leading review of water quality data has concluded that limited data on stream quality show no clear national improvement in water quality.[22] Some water bodies have become cleaner, thanks to improved sewage treatment and controls of point sources, but others have become dirtier, usually because of uncontrolled non-point source pollution.[23]

Wetlands regulation likewise reflects the failure of the regulatory system to address decentralized threats adequately. The United States lost more than 50 percent of the wetlands of the contiguous United States by 1990.[24]

Because wetlands provide habitat for a rich variety of species, clean water, and control floods, the United States has sought to protect wetlands for years. But since wetlands get destroyed piecemeal through thousands of individual land development projects, efforts to protect them have been less than completely effective.

Indeed, land development, a process of numerous project level decisions, many of them private, generally poses a great threat to the survival of many species. But under the Endangered Species Act, the government only rarely stops projects that might threaten habitat.[25]

When people think of environmental problems, they often think of a few bad big businesses wreaking havoc. While some important industries certainly contribute disproportionately to environmental problems, many problems involve numerous, hard to regulate sources.[26] As population and consumption increase around the globe, these diffuse sources will become ever more numerous and important. The developed countries have largely failed to adequately address problems coming from diffuse pollution sources. Developing countries with their larger populations may have even more difficulty, even if their capacity for environmental protection increases as their wealth does.

In the United States, a very large country, the regulatory system has decentralized decision-making, to a degree, to try and address the problem of decentralized sources. Thus, the Clean Air Act requires states to develop plans to address air pollution problems stemming from numerous sources,[27] and the Federal Water Pollution Control Act relies upon state standard setting to address non-point pollution sources.[28] But the economic and political dynamics of federalism have limited the effectiveness of this decentralization. States have done almost nothing about non-source water pollution for three decades and have rarely met the goals set out the Clean Air Act for state implementation plans.[29]

State pollution control officials regularly claim that fears about competition from unregulated (or underregulated) out of state industries make it difficult for states to vigorously regulate their own sources.[30] Professor Revesz of New York University School of Law has challenged the claim that a "race to the bottom" among states justifies a strong federal role.[31] But one does not have to believe in a thorough-going race to the basement to believe that economic competition between states might help explain the recurring failures of state regulation. And Kirstin Engel of Tulane Law School, a critic of Revesz's arguments, has pointed to empirical evidence demonstrating that economic competition influences decisions to weaken state programs.[32] She also shows that Revesz does not dispute the idea that economic fears may retard state regulation.[33] Rather, he argues that some retardation may be optimal.[34] So his argument is best understood as another efficiency argument, a dispute with the policy choices Congress has made for the environmental system.[35] Futhermore, Professor Engel explains that the insights of game theory support the conclusion that inter-state competition may interfere with efforts to achieve optimal pollution levels, since states may act out of fear that their neighbors will set weak standards to lure away some of the fearful state's industry.[36]

In any case, delegation to states has not solved the problem of responding to pollution problems with numerous sources. Delegation to states, even if effective, would not imply as much decentralization as occurs in making private decisions to harm the environment. And not even Revesz claims that state regulation has effectively met the goals that Congress has established for the regulatory system.

I do not claim that the millions of private decisions that might destroy the environment invariably do so. For example, while decisions to fell trees have destroyed especially valuable old growth forests, overall forest cover has grown in the United States. Eastern forests have actually increased in extent in recent years. But I do claim that a serious issue exists about how the slow, cumbersome regulatory process we have can keep up with the substantial environmental threats from numerous sources of pollution.

New Concerns

So far, the analysis has focused upon old and well-recognized problems that the regulatory system has aimed to solve for several decades. But the bigger issue in an economy that grows and changes over time may be the development of new problems. Adaptive efficiency would seem to demand large numbers of creative experiments to address these newer problems. Indeed, as we have seen, the free market encourages innovation even in the face of substantial uncertainty. This is especially true when the product has a lucrative potential market. In other words, if a product *might* deliver a great material benefit, it will often be introduced even though that benefit is uncertain.

The regulatory system, however, does not typically foster bold experiments when facing mere possibilities (or even fairly reasonable likelihood) of future threats. While rarely extremely vigorous, the regulatory system can become dangerously somnambulant when facing something new. Climate change provides a striking example of this. Several gases emitted into the atmosphere trap heat in the lower atmosphere, thus warming the earth's surface. Because of past emissions of these "greenhouse" gases, the earth's mean surface temperature has risen significantly and scientists predict more warming in the future.[37] While almost no mainstream scientists doubt these facts, the precise effects of the predicted warming remain uncertain. Possible consequences include the inundation of coastal areas and island states (from melting polar ice caps), the spread of infectious diseases (as warm climates spread northward), increasingly frequent violent weather events, and ecosystem destruction.[38]

Emissions of greenhouse gases have risen throughout the industrial age and continue to rise today. These emissions are rising at an enormous pace in developing countries. Atmospheric concentrations of carbon dioxide rose 13 percent in the two centuries before 1960.[39] Between 1959 and 1998, a period of just thirty-nine years, carbon dioxide levels increased another 17 percent. The twenty-nine industrialized countries in the Organization for Economic Cooperation and Development have seen a 9 percent increase in carbon dioxide emissions just since 1990.[40]

Climate change provides a global example of a problem stemming from enormous numbers of diffuse pollution sources, some of which are very difficult to regulate. The sources of this problem span the globe. Any burning of fossil fuels produces carbon dioxide, a major greenhouse gas. That means that driving a car, running a power plant, or operating a factory anywhere in the world contributes to global warming. Methane emissions from gas pipelines, some types of gasoline burning, landfills, and livestock operations add to the warming. Nitrogen oxides, a product of incomplete fuel combustion (principally from automobiles and power plants), also contribute to climate change. Other gases contribute as well.

And the gases tell only half the story. The earth's oceans and terrestrial ecosystems ameliorate climate change by sequestering carbon. Therefore, the destruction of ecosystems contributes to climate change. For example, burning a rainforest both produces carbon emissions and lowers the sequestration capacity of the rainforest. This means that peasants clearing land for agriculture, cattle ranchers burning forests to establish ranches, and loggers cutting trees in order to sell wood (for example) contribute to climate change. Nearly every human activity has some effect on climate change.

The regulatory system's economic dynamic failure here, however, involves more than just the problem of a slow system struggling to address a problem having numerous sources. Rather, it reflects the extremely conservative nature of the system—the difficulty our centralized political system has in coming to grips with new information. Whereas the system encouraging adaptively efficient material innovation treats changed conditions as an opportunity to take risks and reap rewards, the regulatory system often regards any uncertainty at all as a reason to do nothing. Indeed, even when the potential problems are extraordinarily serious and, therefore, the potential value of action very high, the regulatory system does little to foster deployment of innovations that would address the problem.

The world has struggled to address climate change for over a decade. It took nearly ten years to agree to any concrete environmental protection at all, in the Kyoto Protocol to the Framework Convention on Climate Change.[41] Many scientists consider the measures agreed to at Kyoto inadequate.[42] And the United States' decision not to ratify the protocol has greatly weakened it.

The irreversibility of some environmental problems, including climate change, would suggest that adaptive efficiency requires a vigorous response to pollution that accumulates in the environment over time. Recall that adaptive efficiency aims to preserve the maximum number of future options. If we discover that an irreversible problem has more serious consequences than we thought, we have no way of revisiting our decision to allow the status quo to continue. We cut off future options to avoid environmental damage when we fail to act in a timely manner.

Climate change involves irreversible consequences, because greenhouse gases remain in the atmosphere for a very long time. This means that when we allow greenhouse gas emissions to continue, we must live with the environmental consequences of that decision for many decades, even if we subsequently reverse the decision.

We narrowly and only partially escaped very serious problems with ozone depleting chemicals, which also have long residence times. Because we banned the most important ozone depleting chemicals, scientists have predicted that the current hole in the ozone layer, which we acted too late to prevent, will mend in the coming decades. Scientists predict that skin cancer will increase before the mending is complete, but vigorous, timely action prevented more widespread consequences.[43]

Persistent toxic pollutants likewise pose risk of irreversible problems, including potentially serious threats to reproduction. As a result, a number of nations have agreed to ban a small list of persistent organic pollutants.[44] This list does not include many potentially harmful substances that bio-accumulate.

Since a global solution to the irreversible problem of climate change will probably require development of alternatives to our fossil fuel based economy, the need for innovation here is fairly obvious. A failure to develop and diffuse the innovations needed for growing economies around the world to enjoy adequate energy without exacerbating climate change may prove extremely serious. Countries like China, India, and Brazil are expected to become far more industrial in the coming decades than they are today. Since these countries have very large populations, an increase in industrialization in these countries based on existing fossil fuel technologies would greatly increase greenhouse gas emissions—increases that will add to levels scientists already consider unsafe. Unless these countries forego much needed economic growth, which seems both inequitable and unlikely, they must have newer, cleaner technologies to make that growth compatible with the climate. Absent radical innovation, increased future emissions will combine with past pollution remaining in the atmosphere to exacerbate changes in climate.

From the standpoint of efficiency, climate change is a very controversial issue. Since the science cannot pinpoint the precise regional consequences of climate change, quantifying benefits involves a lot of guesswork. Similarly, since predicting future technological change is problematic, predicting the costs is difficult as well. Finally, no one has come up with a really good method for comparing costs to benefits even when they are known, because of the difficulty of putting a monetary value on human life and ecosystem destruction.[45] From the standpoint of adaptive efficiency, the case for moving on climate change seems much less difficult to analyze.

How the Shape of Environmental Problems Should Refocus Debate

These problems cumulatively point to a concern about changing environmental quality over time. They raise questions about whether our regulatory system will produce enough timely decisions to meet the challenges of ever-changing, and often decentralized, generation of environmental problems.

This rate of change concern may be much more important in the long run than concerns about the efficiency of any particular regulatory decision. A focus upon rates of change over time differs markedly from the question usually asked about government regulation. Most analysts ask, "How can we make better regulatory decisions?" The analyst critiques past government regulatory decisions and derives recommendations for reform from that critique. The analyst's implicit or explicit normative views about what good administrative decision-making consists of heavily influence both the critique and the reform recommendations. These days, as part I suggested, that kind of analysis tends to focus on each regulatory decision as a potentially inefficient use of private compliance resources, and usually recommends efficiency enhancing reforms. Focusing upon the relative rates of change in the environment involves a more systematic and broader approach. Analysis of relative rates of change focuses upon the regulatory system not as a series of transactions but as one of several influences upon the environment. It concerns itself with a question of adaptive efficiency, whether the regulatory system can adapt to changing conditions around it, not with administrative decision-making as an isolated phenomenon.

If the government regulated everything, then the question of whether regulators made good decisions about the particular matters before them would provide an adequate basis for assessing whether existing systems will adequately protect the environment over time. But regulators do not and cannot regulate everything that affects environmental quality. If the regulatory system makes perfect decisions every time, it does not follow that environmental quality will be adequate, if most of the sources of problems simply receive no attention. This simple fact raises profound questions about how environmental law might keep up. And those questions might include thinking about altering the economic dynamics that limit environmental innovation and virtually cut off radical environmental innovation.

This part has described the economic dynamics of environmental law. It developed a theory about the value of innovation, with separate quantitative and qualitative components. It explained why economic incentives might favor innovations damaging the environment over time. It explained why government decision-makers facing well-financed pressure to do little and enjoying no economic incentive to induce innovation might often fail to induce environmental innovation, while decentralized private decision-makers have significant incentives to make numerous decisions to facilitate introduction of material innovations, some of which are bound to be environmentally destructive.

Finally, this chapter provides a foundation for explaining more broadly why the slovenly pace of government decision-making relative to competing private decision-making may prove inadequate to many environmental challenges. First, private decisions can multiply the significance of small, hard to regulate causes of environmental problems. Second, when facing new challenges, the government decision-making often tends to retreat, rather than to experiment with solutions. This analysis of the economic dynamic influencing environmental law leads to questions about how to improve environmental law's economic dynamics. If we cannot really expect a slow regulatory system to keep up with a raft of decentralized private decisions trying to meet the desires and needs of a growing population wishing to increase consumption, perhaps the economic dynamics need to change. The next part presents some of the questions that economic dynamic analysis invites about how to change environmental law to improve economic dynamics.

III

Toward Economic Dynamic Reform

This book began by suggesting that perhaps environmental law should mimic the real economic dynamics, rather than the hypothetical efficiency, of free markets. After showing some of the problems with efficiency-based approaches in part I, part II compared the economic dynamics of environmental law to the dynamics of the free market. It suggested that environmental law's economic dynamics might prove inadequate in the future, encouraging insufficient innovation to counteract the predictable growth in threats to the environment.

Understanding of the economic dynamics of environmental law and the importance of that dynamic to our future should force us to change the questions we ask about reforming environmental law. Most analysis of environmental law these days focuses upon improving its theoretical efficiency. But this may not be the most important question to ask, and some of the reforms that efficiency analysts recommend would make environmental law even more moribund than it is now.

The next part illustrates how an emphasis upon economic dynamics should change the questions we ask about how to reform environmental law. Rather than ask how we can make environmental law more efficiently use private sector resources (like the hypothetical free market), we could ask how we can make environmental law more dynamic, like the actual free market. This would, in turn, lead us to ask whether it is possible to reform environmental law to make its dynamics more closely resemble the dynamics of the free market. The free market offers relatively rapid decentralized private decision-making, which fosters a willingness to experiment. This invites inquiry into whether we can and should privatize environmental law. If privatization is either undesirable or too difficult to accomplish completely, we might ask what can be done about the slovenly pace of government decision-making. How can government make reasonably numerous and rapid decisions, just as entrepreneurs and some businesses do as they innovate to improve our material welfare? Finally, how can one improve the design of environmental regulation to encourage innovation and sustainable development? Analysts have been asking about how to improve the efficiency of regulatory design, but focus little or no attention upon sustainable development and address innovation only as an incidental byproduct of efficiency.

As this part explores these questions—the questions of privatization, increasing the scope and speed of government decision-making, and improving regulatory design—it will also suggest a number of possible reforms. Some of the reforms suggested are quite radical and receive only a limited defense here. My goal is not to persuade the reader that we should adopt all of the reforms I mention. Rather, I use these reforms to make the questions the reforms address more concrete and understandable. I hope that this discussion persuades the reader of the importance of the questions these reforms address. Similarly, the discussion sometimes frames questions for further research. I raise these research questions to show that reorienting analysis toward the questions I focus upon raises some interesting questions that have received insufficient attention. If the reader agrees about the importance of the questions I raise, this book will have succeeded, even if the reader comes to different conclusions than I suggest about what reforms would best address the questions raised.

9

Privatizing Environmental Law

Since the private market provides a more economically dynamic environment than the regulatory system, one might ask whether government could privatize environmental law. While privatization includes some practices and reforms that have already been implemented or involve substantial continuity with some past practice, it might also involve some extraordinary legal innovations that may strike some readers as impracticable. Raising this question of privatization, however, helps to show how institution bound our current thinking is. We take for granted an administrative law structure that performs badly from an economic dynamic perspective, without even asking whether the institutional arrangement might be radically changed to address this problem. I hope that readers at least will see that we must not take these inadequacies for granted and should think seriously about addressing them. Subsequent chapters offer some more modest proposals for reform.

The question of how to privatize environmental law embodies a conceptual difficulty worth highlighting at the outset, the questionable distinction between private and public realms. Our private markets do not consist of some free state of nature. Rather, they reflect and depend upon legal rules that the state establishes. So the private market is not really private in the sense of having no dependence on government and law.

If the privatization question asks whether one could have environmental protection without some law encouraging it, the answer is no. As the previous part explained, the private market does not provide incentives for even very worthwhile environmental protection if left to its own devices. Some intervention is needed.

Nevertheless, governments can design environmental law to greatly improve its economic dynamics. Legislative bodies could privatize—leave to nongovernment decision-makers—more of environmental law's functions in ways that make it much more dynamic. The law could, for example, avoid using agency standard setting as the principal mechanism to meet environmental objectives. This might lead to more frequent and decentralized standard setting, which might help environmental law keep pace with private decision-making. This decentralized approach might spur more innovation than the existing plodding regulatory approach.

While privatization of environmental law seems like a new idea, some aspects of it have been privatized already. By analyzing more precisely the actual economic dynamics at work in existing privatization, one can better appreciate the nature, strengths, and limitations of the privatization that has occurred.

This leads to some ideas for improving the economic dynamics both by enhancing the economic incentives that already encourage some privatization and by considering new forms of privatization. My main point is that an important set of issues emerges from considering the privatization question. The proper resolution of these issues may enhance the dynamism of environmental law.

Privatized Enforcement

The United States has partially privatized enforcement of environmental law. Citizen suit provisions in environmental statutes provide for private enforcement of government set pollution limits. Congress recognized that government enforcement might be erratic. This reflects, in part, an understanding of economic dynamics. Just as industries can use their resources to discourage strict standard setting, they can use their resources to pay people to dissuade the government from vigorous enforcement.

Furthermore, even a government willing to stand up to anti-enforcement pressure lacks the resources to enforce the law every time a major violation occurs. Citizen suits augment government enforcement resources.

Congress created an economic incentive to bring citizen suits by providing for awards of attorney fees to victorious plaintiffs. This incentive has made it possible for attorneys who have some other sources of funding to bring citizen suits in the hopes of accomplishing environmental goals. The possibility of citizen suits no doubt deters many violations of environmental standards.

This deterrent proved especially important during the Reagan administration. That administration simply disagreed with many of the environmental laws Congress had passed, so it often declined to enforce them. The Natural Resources Defense Council (NRDC), a citizens' environmental group, collected data indicating a large number of clear violations of Clean Water Act standards. When the federal government declined to enforce the law, NRDC and other groups brought citizen suits to enforce it. In 1983, NRDC and the New Jersey Public Interest Research Group, groups with few lawyers involved in enforcement, brought almost as many enforcement actions as the entire EPA.[1] For a period in the 1980s citizen suits became the dominant means of enforcing the Clean Water Act.

Yet, the incentive of attorney fees for victorious plaintiffs does not, by itself, make private enforcement of environmental violations economically attractive enough to secure widespread private enforcement of environmental law. The law has become so complicated that plaintiffs' lawyers cannot quickly sort winning cases from losers, and may lose cases involving serious violations of environmental laws.

Citizen suits against polluters are complicated, because environmental standards are often complex and good monitoring data are sometimes hard to obtain and evaluate. A conscientious attorney must carefully check the monitoring data against the standards before bringing suit.[2] If she finds that no violation has occurred, she will not bring suit and will receive no compensation for the time spent finding and analyzing the law and the data.

Even if she does find a violation, she still may lose her case in court. For one thing, the Supreme Court has disallowed citizen suits for wholly past violations under some statutes.[3] The citizen suit provisions in the environmental statutes require plaintiffs to give notice before bringing suit. If the defending polluter fully cures the violation before the plaintiffs can file suit, she may receive no compensation for her time under some statutes.

Even where a statute authorizes suits for wholly past violations, the Court has created uncertainty about whether the Constitution always allows suits when a polluter cures a violation prior to filing. In one case, *Steel Co. v. Citizens for a Better Environment,*[4] a citizens' group sued a company for civil penalties for violating pollution reporting requirements, but the company filed the necessary reports after receiving notice of the suit before the group filed suit in court. The Supreme Court held that the civil penalties and other requested relief would not redress the violation. Because of this, said the Court, consideration of this case would violate the doctrine of "standing," judge-made law specifying the meaning of the constitutional provision allowing courts to hear cases or controversies. The Court has held that no case or controversy exists when standing to sue does not exist.

In a subsequent case involving violation of actual pollution control requirements, the Court held that civil penalties can redress violations and reversed a lower court decision holding that the standing doctrine precluded suits when a violator cures after filing.[5] Since the Court did not specifically overrule *Steel Co.*, uncertainty exists about when suits for violations cured before filing will be allowed under the Constitution. The case also left open the possibility that in some cases cure after the violation might make a case moot and therefore non-justiciable.[6] All of this means that a citizen suit for a clear violation of the law can sometimes fail in a way quite costly to private enforcers.

If the defendant does not cure the violation, some plaintiffs may file suit and pursue it to judgment. But such plaintiffs receive no compensation prior to potentially long, expensive, and arduous litigation. So, the lawyer representing indigent clients must finance the litigation herself prior to the fee award. Professor William H. Rodgers, Jr. of the University of Washington School of Law has estimated that the requirements of showing standing alone, a prerequisite for even addressing the merits, generated $125 million in added litigation costs in the 1980s, a substantial share of which initially comes out of the plaintiff's pocket.[7] If the plaintiff's lawyer loses, she receives no fee award (even if the issue is close). If she wins, she may be compensated for the case won, but not for other cases that were researched and proved wanting or for cases where defendant won because of a jurisdictional defect (e.g., standing or mootness).

This means that a lawyer thinking about living off of citizen suits will not receive compensation for all of her time, even if she is very successful. Good lawyers litigating complicated cases in other areas generally receive much more attractive financial incentives. Lawyers representing wealthy clients get paid whether or not they win their cases.[8] Private lawyers representing indigent plaintiffs in complicated actions for damages cannot be sure of payment, so they often take cases with very high potential damage awards and demand a significant "contingent fee," that is, a promise of a large percentage of the client's recovery if they win. This compensates them for the risk of losing and getting nothing in at least some of their cases.

The economic dynamics limit the enforcement of environmental laws. Analyzing the question economic dynamics highlights, how the precise incentives interact with the bounded rationality of the person subject to the incentives, reveals the problems. In this case, citizen suit provisions are supposed to provide incentives for lawyers to represent citizens seeking pollution abatement, so we must analyze the incentives the law and the private market offer these lawyers. Lawyers capable of handling complex cases receive economic inducements to prefer other areas of the law, where compensation is either certain or very high in the event of victory. As a result, often no lawyer will be available to help detect and enforce violations of environmental laws. This shortage may especially handicap poor communities that wish to use citizen suits to reduce environmental burdens, but lack the resources to pay lawyers.[9]

One could correct this. Currently, civil penalties for violations go into the United States Treasury, even if a citizen suit provides the vehicle for collecting the civil penalties. Perhaps these penalties should go to the winners of citizen suits and their attorneys. Alternatively, Congress could reverse *Burlington v. Dague*,[10] a Supreme Court decision precluding high fee awards for especially complicated and risky environmental cases.[11] Justice Scalia's opinion for the Court reflected a concern that a fee raised to reflect the case's risk and complexities would make the incentives to bring nonmeritorious claims equal to the incentives to bring meritorious claims. This view is clearly incorrect. Since plaintiffs receive nothing at all if they do not win under the citizen suit provisions of modern environmental statutes, they always have an incentive to choose meritorious claims and decline nonmeritorious cases.[12] They are more likely to win when they have a meritorious claim.

The efficiency-based concept of "optimal enforcement" offers a potential challenge to efforts to improve the economic dynamics of citizen suits (or any other effort to improve the effectiveness of enforcement).[13] This concept posits that government should design enforcement policy to provide for an optimal level of lawbreaking. Scholars adopting this point of view share a concern for "overdeterrence," the problem of high penalties or very effective enforcement deterring socially desirable conduct.[14] For example, high damage awards in medical malpractice cases may cause doctors to waste a lot of money and time on unnecessary lab tests, i.e., practice defensive medicine.

In the environmental area, the law and economics literature supports a fine equal to the value of the pollution's harm, inflated by the chances of detection, plus the variable enforcement cost of imposing the fine.[15] The legal system has generally not sought to calibrate fines in this manner and, in any event, judges, prosecutors, and lawmakers cannot calculate the value of these variables with even reasonable accuracy. Determining the "chances" of detection alone would require a huge amount of data about how many people are getting away with environmental violations. But people get away with violations of the law precisely because developing comprehensive data on concealed violations is impossible.

At bottom, adoption of this optimal enforcement ideal assumes that some conduct that environmental laws prohibit should be allowed, and uses weak enforcement to accomplish that goal. Thus, Michael S. Greve, a proponent of optimal enforcement, recognizes that environmental laws are underenforced relative to the enforcement needed to realize their goals.[16] He nevertheless criticizes citizen suits as likely to promote overdeterrence, by which he means compliance with laws not calibrated to match costs and benefits.[17] The concept of optimal enforcement's emphasis on fines proportionate to the harm caused by non-compliance implies rejection of fines adequate to fully enforce laws not based on cost-benefit analysis. The concept thus reinforces the efficiency-based regulatory reform agenda by treating full enforcement of laws not based on such analysis as illegitimate.

Citizen suits generally can succeed only when a polluter has committed a violation of the law. This means that Congress or the presidentially supervised agencies to which it delegates authority has already determined that the conduct producing a successful enforcement action is socially undesirable. It would seem that the chief concern at the enforcement stage should be effective enforcement, not optimal enforcement. If the criteria used to determine the legal standards are inadequate, that should usually be corrected at the law writing stage. In fact, enforcement is not only inadequate to achieve the statute's goals, it is currently inadequate to assure compliance with the regulations that only partially implement these goals. More consistent deterrence of violations of pollution standards is desirable, and privatization aids achievement of that goal.[18] Because citizen enforcers would have even greater incentives to bring cases involving gross violations than they do now if they received the civil penalties (since higher penalties would come about in such cases), such a reform would further encourage greater attention to serious violations. Economic dynamic reform might help achieve more widespread compliance with environmental law by strengthening citizen suits.

This example shows that privatization, while sometimes viewed as a recent conservative legal reform, has a history of enhancing the dynamism of environmental law. Furthermore, it shows that an economic dynamic analysis, focusing upon how incentives interact with the bounded rationality of those it hopes to motivate, contributes to enhancing privatizing reforms' efficacy in bringing more dynamism to enforcement. The efficiency-based alternative, in this case optimal enforcement, would serve to make court proceedings more uncertain and the enforcement system more moribund.

Privatized Standard Setting

Citizen suits against polluters, however, only enforce pollution limits that the government has established through its regulatory program. And the economic dynamic analysis in part II identifies the slovenly pace of standard setting as a major economic dynamics problem. One might ask whether one can privatize standard setting.

Information Strategies

We have some experience with privatized standard setting in response to laws generating information about pollution and the cost savings available from some pollution reducing acts. This experience provides further evidence of the potential of privatization to bring fairly vigorous responses from lots of decentralized decision-makers able to make environmentally favorable changes. An analysis of this experience shows that an important feature of efficiency-based analysis, a perfect information assumption, obscures some basic research questions that this experience invites. Again, an interest in the interaction between inducements for action and the bounded rationality of actors subject to the inducement (in this case factory owners making disclosure) raises some good questions for research. And good answers to these questions might aid desirable regulatory reform.[19]

Information about Pollution Levels For many years, polluters in the United States only revealed their pollution levels in response to sporadic specific government requests, usually as a prelude to a regulatory proceeding. In 1986, however, Congress began requiring large manufacturing companies to regularly report discharges of more than three hundred toxic pollutants and EPA created a toxics release inventory (TRI) from this information. Many companies asked their employees and consultants, often for the first time, to estimate factory pollution levels for the listed substances. The amount of pollution found and reported, (10.8 billion pounds in 1987), shocked the public, EPA, and even the companies themselves.[20] EPA had assumed that pollution levels were far lower than they turned out to be and so, apparently, had the management of many reporting companies.

Companies reporting these high pollution levels often found the press and the public responded most unfavorably. And many companies began to look for opportunities to prevent some of this pollution.[21] Some did so under considerable pressure from concerned citizens living near polluting plants. One activist group carried out a hunger strike to protest high pollution levels at a neighborhood plant. The first Bush administration subsequently secured voluntary commitments from several large companies to reduce emissions of some high priority toxic pollutants.

Many other examples of information disclosure requirements inducing voluntary pollution control could be cited.[22] Some companies have reduced or eliminated use of toxic chemicals pollution in response to a California law requiring products containing carcinogens to have warning labels advising consumers that the product might cause cancer.[23]

The academic literature often discusses information strategies as "economic incentive" programs.[24] But the general literature often says little about why or how information provides an economic incentive. This failure reflects the custom of assuming that economic incentives are pervasive and important, without thinking through how they actually work and what other forces may influence people. One must ask why companies sometimes reduce pollution, even in the absence of government regulations requiring the reductions, to begin to think about the economic dynamics of information. In other words, one must, as this book has done, consider both the free market incentives and the effects of government incentives together upon the bounded rationality of private managers. So far, however, empirical work on this question has been very limited and generated inconclusive results.[25]

Many experienced environmental lawyers believe that companies' managers care about their company's reputation.[26] This hypothesis, however, invites further analysis. Perhaps the managers care about a company's reputation because the public might think poorly of them if they report high pollution levels and people might respond by not buying their products.[27] This would suggest that information affects consumer behavior and therefore forms an economic incentive for the company to set standards for itself. The companies respond to a fear of revenue loss.[28] Or managers may fear that investors will decline to purchase a company's stock if a company acquires a reputation as a big polluter.[29] This might deprive the company of capital that could help the company conduct its operations. These stock value and purchase decision theories offer an economic dynamic explanation of information's use, for they explain precisely how information might create an economic dynamic that encourages private cleanup.

Another possibility exists, however. The managers may simply value the company's good name in the community for its own sake. Managers may derive some of their self-esteem from serving a "good company." They may therefore wish to avoid problems that make their company look bad. This may induce them to reduce pollution so that the company can look good and they can feel good about themselves. Notice, however, that this shame and esteem hypothesis does not support the notion that information strategies act as "economic" incentives to induce pollution reduction. Rather, it suggests that information provides a non-economic incentive for reductions based on a desire for good standing in a community.

Perhaps reputation has little to do with some companies' actions. Some managers may have some genuine concern for public health and the environment, and the confrontation with facts makes it difficult for them to do nothing in good conscience. This would imply that information creates ethical or moral, not economic, incentives to reduce.

More cynically, one might think that the companies act to ward off potential regulation. Indeed, TRI data did strengthen efforts to regulate toxic pollution more extensively.[30] This might suggest an economic dynamic. Perhaps the companies believed that government regulation would involve more expense than voluntary standard setting.

In any case, informational strategies sometimes motivate private standard setting. But informational strategies have limits. For one thing, the public can only absorb a limited amount of information.[31] This matters if we expect a strategy to create an economic dynamic by motivating consumer or investor behavior. An information strategy that relies in part on fear of regulation will only work if some political will exists to regulate. While some companies have responded positively to informational incentives, some do not respond at all. Tobacco companies, for example, have terrible reputations and help cause many horrible deaths annually. Yet, they continue to sell cigarettes. So, apparently, some limits exist to the potential of information to induce improvements out of a general non-economic concern for reputation or a desire to protect public health.

Another important limit to voluntary responses to information exists. Voluntary responses typically (although not always) involve a search for cheap, nondisruptive environmental improvements, not radical, new environmental technology.[32] Companies responding to the TRI data undertook many worthwhile pollution prevention projects. These projects often involved some redesign of a process in order to produce less pollution. But they did not involve, for example, displacing pesticides with organic farming. Farming by these techniques would reduce pollution levels at factories making pesticides and around farms now using pesticides far more than the many worthwhile projects voluntarily undertaken in response to TRI data.

This observation flows from an understanding of bounded rationality. New information may make a company perceive pollution prevention to be adaptively efficient. But it is less likely to lead a company to act in ways that damage its core business in a fundamental way. Information about harmful chemicals in food, however, might influence consumers, since they are not bound by the consideration of not damaging a particular producer.

Information about Profitable Pollution Prevention   A second type of information strategy involves disseminating information about the profits sometimes available through pollution prevention. Recall that an economic dynamic analysis predicts that companies would voluntarily implement some profitable pollution prevention projects. But firms often have difficulty seeing these economic opportunities. Firms generally employ people hired because of their skills in helping the company make its product or offer its services. These people are generally quite busy trying to do what the firm usually does. Absent some institutional change, few firms see and seize environmental opportunities.[33]

Firms, however, change over time. Environmental regulation has led many companies to hire employees with environmental expertise. Many large manufacturing firms now have vice-presidents assigned to address issues of occupational health and safety and environmental protection. Some companies have made pollution prevention a focus of significant corporate endeavor. Hence, the general shape and direction of environmental law affects the extent of corporate capacity to seize environmental opportunities.

Some firms work with outside parties to spot opportunities that their own employees cannot see. McDonald's, for example, took a host of environmental measures after consulting with the Environmental Defense Fund (since renamed Environmental Defense), a citizens' group. Perhaps one of the most interesting cases involves Dow Chemical. Dow has a substantial staff devoted to environmental protection issues. Yet, the NRDC helped Dow identify profitable pollution prevention opportunities that their own employees had not found on their own.[34] This is a little surprising. Because Dow's staff knows its own processes better than the NRDC does, one would not expect Dow to profit much from outside help. This implies, however, that even very sophisticated companies find it difficult to get complete information about money saving pollution prevention opportunities.

This information deficit implies that government can sometimes secure voluntary pollution prevention by disseminating information about the cost savings available from pollution prevention projects. EPA has pursued this strategy through its Green Lights Program. It informs industries about the cost savings available through purchase of initially expensive lighting that conserves energy, lowers electricity costs, and thereby reduces emissions associated with power generation. Often these projects pay for themselves in a few years and generate profits thereafter. This program has proved enormously successful.[35]

Scholars refer to this kind of information strategy as an economic incentive program as well, again without analyzing what exactly this has to do with economic incentives. Careful consideration of the economic dynamics, i.e., the interaction of private market conditions and government incentives, clarifies what happens in these programs. Notice that the government does not create any economic incentives to reduce pollution in this program. The incentives existed all along. With or without the government intervention, a profitable opportunity existed to conserve energy through use of energy efficient lighting. The government, instead, helps industry to see the opportunity already present in the economy.

Perfect economy efficiency models actually obscure our ability to analyze this kind of information program. In an economy with "perfect information," polluters would already know about the opportunities to save money while improving the environment. They often do not know about these opportunities precisely because they lack perfect information, and because many environmental problems appear peripheral to managers trying to figure out how to efficiently produce a product. Government can remedy the information deficit because it has an entity, the EPA, that makes environmental improvement its primary business.

Government can provide some kind of incentive for private actions improving the environment, perhaps an economic one, by providing information about environmental problems. It can also motivate private changes by providing information about existing economic incentives for profitable pollution reduction strategies that busy managers may overlook.

Private Environmental Certification Programs
Some industries have engaged in cooperative non-governmental standard setting ventures. Examples include the chemical industry's Responsible Care Program, the International Organization for Standardization's ISO 14000 environmental management program, and the Forest Stewardship Council's well-managed forests program.[36] While these programs are too new to evaluate fully, private standard setting has some capacity to promote environmental protection.[37] Reason exists, however, to doubt that such processes

will spur rapid innovation. Consensual standard setting by existing industry just seems like the wrong process for doing that, since radical innovation may threaten existing industry.

On the other hand, private standard setting may provide for some environmental progress, especially in areas where the problems outstrip the management capacity of government.[38] For example, some environmental organizations have participated actively in the Forest Stewardship Council, because government has proven so ineffectual at promoting sustainable forestry.

All of these examples of private standard setting illustrate that a public action inducing private standard setting can induce the kind of decentralized voluntary response that we associate with the free market. An economic dynamic analysis yields a more precise description of these programs, including both their virtues and limitations, and identifies some areas for further research that might aid better design of informational programs.

An Environmental Competition Statute

So far, this chapter establishes that existing environmental law privatizes some enforcement and standard setting. It also shows that economic dynamic analysis might help aid understanding and study of informational strategies and lead to productive improvements in existing privatization mechanisms.

Focusing upon privatization as a possible reform, however, should raise questions about whether privatization might go further than it has in the past. The economic dynamic analysis leads to questioning the centralized decision-making structure that pervades much of environmental law, and to asking whether we can supersede this relatively moribund mechanism and create something that more closely resembles the economic structure of free markets. Securing maximum incentives for innovation may require legal structures that induce competition to produce environmental improvement and lessen the need for repeated government decisions. A more dynamic system might seek to decentralize standard setting and make it more likely to promote innovation.

Pollution taxes may create continuous economic incentives to reduce emissions, but they do not rely on the dynamic that drives a competitive free market—competition among firms. Rather, the incentive comes from the same source as incentives in traditional regulation—government decisions. Similarly, emissions trading relies upon government set emission limits to drive demand. Hence, these mechanisms suffer from some of the same structural defects that limit the dynamism of traditional regulations, namely, dependence upon plodding government decision-making, rather than ever-improving private capacity, to set goals.

These efficiency-based reforms do not seek to emulate the dynamics of a free market. Rather, they seek to emulate its hypothetical efficiency. In a free market, companies must innovate in response to competition, not in response to government decisions about taxation and permitting. Imitating that dynamic is very different from just affording polluters flexibility in how they respond to government decisions.

We can design more dynamic economic incentives that encourage competition to reduce pollution, much as the free market creates competition to provide better amenities. This requires creation of mechanisms that circumvent the need for repeated government decisions and allow private actions, rather than government decisions, to stimulate reductions in pollution.

The law can apply either positive economic incentives, such as revenue increases or cost decreases, or negative economic incentives, such as revenue decreases or cost increases, to polluters. This reveals a possibility that has received too little attention. Negative economic incentives can fund positive economic incentives.

Governments have designed programs that use negative economic incentives to fund positive economic incentives. New Zealand addressed the depletion of its fishery by imposing fees on fishing, a negative economic incentive, and using revenue from these fees to pay some fishermen to retire, a positive economic incentive.[39] This may reduce pressure on the fish if fees are high enough. The California Legislature has considered a program, Drive +, that imposes a fee upon consumers purchasing an energy inefficient or high pollution vehicle.[40] The proceeds would fund a rebate on the purchase of an energy efficient vehicle or low polluting vehicle.[41] Similarly, New Hampshire officials have proposed an "Industry Average Performance System" that would redistribute pollution taxes on the polluting industry in ways that favor lower emissions.

One can build on this principle of having negative economic incentives fund positive economic incentives to craft laws that mimic the free market's dynamic competitive character far better than taxes or subsidies. In a competitive free market, a firm that innovates to reduce its expenses or increase its revenues not only increases its profits, it often reduces its competitors' profits. Hence, firms in a very competitive market face strong incentives to innovate and improve. Failing to innovate and improve can threaten their survival. Implementing innovations and improvements can help firms prosper in a competitive market. One might seek to design environmental law to create a similar dynamic.

One could craft, for example, an "environmental competition statute" requiring polluters with relatively high pollution levels to pay any costs that competitors incur in realizing lower pollution levels, plus a substantial premium, thereby creating a significant incentive to be among the first to eliminate or drastically reduce targeted pollutants. Such a law would simply authorize any polluter to collect costs plus a premium from any competing firm with higher pollution levels. Thus, for example, a power plant that switched fuels to achieve a lower emissions rate per kilowatt hour than its competitors might collect the cost of the fuel switching from its coal-burning competitor, plus a premium.

An environmental competition statute directly attacks a fundamental problem with existing free market incentives—the polluting firm must absorb any cleanup costs. Because the firm does not bear all of the costs of pollution itself (most are externalized and felt by the general public), it rarely pays to clean up. If firms could systematically externalize the costs of cleanup without substantial administrative intervention, just as they externalize the cost of pollution, then even a fairly modest premium might create adequate incentives to control pollution.

This solves another problem as well. The free market system provides no systematic incentive for environmentally superior performance. The environmental competition statute regularly rewards superior environmental performance.

An environmental competition statute would create a private environmental law with a few public decisions setting up the law, but substantial enforcement by low polluting businesses against competitors. The law would create a private right of action that allows a business that realizes environmental improvements through investment in pollution reducing (or low pollution) processes, control devices, products, or services to secure reimbursement for expenses, plus some premium, from more polluting competitors. Hence, the scheme would create economic incentives for some companies to become enforcers of the law, rather than creating incentives for most companies to resist enforcement. This would effectively privatize enforcement, making it a private activity rather than a government activity with some public-spirited private support (as in the citizen suit mechanism).

Such a proposal overcomes the fundamental problems with traditional regulation, emissions trading, and pollution taxes. These mechanisms rely on government decisions as the driver for pollution reductions. An environmental competition law makes private initiative, motivated by the prospect of gain and the fear of loss, the driver of environmental improvement, thus replicating free market dynamics. The magnitude of the incentive may depend upon the extent of industry fears about competitors' achievements, rather than only the limited cost government imposes through regulation (or pollution taxes).

An environmental competition law might seem to create incentives to reduce first and do nothing to motivate reductions from slow movers. However, the dynamic such a program creates, like the dynamic of a free market, works more broadly. No one would know, a priori, who the first movers would be. Thus, anyone who did not actively seek emission reductions would risk financial loss of uncertain dimension; precisely the risk companies face when they fail to innovate in making improvements or new products in a competitive market.[42]

Moreover, such a scheme provides a continuous incentive to reduce pollution. Any company can profit by making an environmental improvement or lose money by failing to make one. The government does need to establish the premium to be paid first movers. But once it establishes this, repeated government decisions are not necessary.

Companies might collude to avoid such a scheme rather than compete to earn money from it. All of the companies subject to the law could defeat it by deciding to do nothing. To prevent this collusion, lawmakers might restrict communication between companies regarding their plans under the law. Communication about reduction plans might be considered a combination in restraint of environmental trade and banned on a kind of antitrust theory. A definition of a competitor broad enough to allow new entrants in markets to compete should also limit opportunities for collusion.

The government might still decide which pollutants this law would apply to. It might make judgments that such strong medicine should apply to some pollutants and not others. Or it might write a very open-ended definition of pollution that allows the scheme to apply to any substance that might have a negative impact on health and the environment. This would give an incentive for first movers to develop information that a substance it planned to reduce has some harmful potential. This would address the problem of the regulatory system not generating enough information about new substances. Currently, the regulatory system provides a disincentive to create information about harmful effects, since such information can lead to regulation, at least when the law requires regulation based on cost-benefit analysis or health and environmental effects.

Like all other schemes, an environmental competition statute would require development of sufficient data to determine pollution levels of facilities. The law would work best if it included some mechanism, such as a requirement that pollution levels be posted regularly on the Internet, that made it possible to see whether a company has performed better than competitors environmentally, without having to obtain information from government files. One could build in an incentive to improve monitoring by providing a strong presumption that pollution levels ascertained through continuous emissions monitoring would be presumed correct, but that claims based on estimation would be discounted in some fashion in determining payments under the statute.

The definition of a competitor from whom an environmentally successful company might claim a payment would play an important role in such a statute. EPA traditionally regulates by grouping industrial processes that share standard industrial classification (SIC) codes and then creating subgroupings to try to address plants with similar environmental or physical characteristics. This makes sense for regulation.

However, SIC codes do not fully describe competitors in a system designed to reward environmentally friendly innovation and apply a negative economic incentive to dirtier means of meeting the same consumer goal. Ideally, someone who develops a system of integrated pest management (IPM), for example, that makes it possible to increase crop yields with little or no pesticides, should be able to collect a payment from pesticide manufacturers that compete with her to maximize crop yields. Even if the IPM developer operates a research farm and the pesticide manufacturer operates a pesticide plant, the statute should regard them as competitors (or allow courts to develop a common law of competition based on broad principles).

The application of an environmental competition statute to a well-defined group of polluters with a very clear specific definition of competitors tailored to one problem would probably not generate large volumes of disputes. For example, EPA could require all electric utilities to pay fees for each ton of nitrogen oxide emissions and divide the proceeds evenly among the five electric utilities with the lowest rate of nitrogen oxide emissions at the end of a year. This would prove rather simple to administer. Broader programs would pose more competitor definition issues, but would offer even broader incentives for transformative innovation and environmental improvement.

Such a law should include a dispute resolution mechanism. Competitor enforcement may produce more need for conflict resolution. An environmental competition law may create commercial disputes resembling those that arise under other commercial laws. Disputes may arise about who is a competitor, what costs a company incurred, and what pollution levels exist at various facilities. One may want to use some fees from polluters to finance specialized arbitration of these disputes.

An environmental competition statute should not generate complicated, environmentally fruitless disputes. The Comprehensive Environmental Response, Compensation and Liability Act of 1980, otherwise known as Superfund, makes a variety of parties associated with toxic waste dumps strictly jointly and severally liable for cleanup. This has often led to protracted disputes largely because apportioning liability among potentially responsible parties (PRPs) has proven difficult.[43]

Superfund, however, has been a notable success in encouraging parties not to create new toxic waste dumps since its enactment in 1980. The Chemical Manufacturers Association reported a decrease in waste generation of 51.8 percent from 1981 to 1985, despite an increase in the value of chemical shipments from $180.5 billion to $214 billion over the same period.[44] An environmental competition law would likely stimulate a comparable scramble to avoid liability by reducing pollution levels.

The principle causes of protracted disputes and high transaction costs under Superfund for already existing toxic waste sites would not exist under an environmental competition statute. Allocating responsibility has proven difficult under Superfund because good information about the past history of toxic waste dumps (who dumped, who allowed dumping, etc.) is hard to obtain and the program creates great uncertainty about the means and scope of eventual cleanup. It usually will not be difficult to determine pollution levels and, accordingly, responsibility under an environmental competition statute, since liability will only arise after a pollution reducing activity is completed and documented and will only attach to a defendant who has not reached the leader's level. Furthermore, transaction costs are highest under Superfund when a toxic waste site involves large numbers of potential parties.[45] One could structure an environmental competition law to limit the parties to as few as two—one defendant and one plaintiff.

PRPs and EPA often seek to allocate responsibility under Superfund before completion of cleanup. This also hinders settlement because the total value of liability remains open-ended at the time of negotiation. An environmental competition statute should only allow claims based on already completed cleanup. Since disputes will only arise after the costs have already been incurred and are known, many potential disputes will be settled.

An environmental competition statute, like any aggressive regulatory program, runs the risk of stimulating unintended environmental consequences. For example, nuclear power plants have low rates of NO_x emissions. This means that an environmental competition statute based on relative NO_x limits might generate payments from coal and gas-fired utilities to new nuclear plants. Unless the government intends to subsidize nuclear power in spite of its generation of radioactive waste and its safety risks, it should write some standards to avoid transfers to companies with technologies that pose great safety risks or generate important alternative pollution problems.

One might object that an environmental competition statute with a broad definition of competitors might allow firms to put their rivals out of business. This, however, raises a question. We typically accept the notion that firms that meet material needs may drive their competitors out of business. We know that this creates job losses. But we accept this consequence anyway. If we accept this consequence in the realm of innovation to improve our material welfare, then no good reason exists not to accept business failures and job losses as part of efforts to improve our environmental welfare.

On the other hand, some countries do not accept competition and employment loss so cavalierly. They seek, in various ways, to protect workers from the dislocation that competition causes. Perhaps we should follow this example. But, if so, we should do so across the board, protecting workers not just from the consequences of changes generating environmental improvements, but also from the consequences of changes improving our material well-being, and certainly from the consequences of mergers, some of which may improve the welfare of no one except egocentric top managers. It's ironic that environmental regulation, which has produced a small net increase in jobs,[46] has come under fire for creating unemployment, while mergers, which account for large numbers of layoffs, receive little governmental scrutiny as an employment issue.

The scheme I have described appears to provide no limits on the cost one firm can impose on another. In this sense, it may appear more dynamic, and less efficient, than the free market. After all, if an innovation in the free market costs too much to produce and no one buys it, then the innovator will fail. But under this scheme, apparently one firm can invest infinite amounts of money in transforming a production process to improve its environmental characteristics and then collect all of that cost from a competitor.

In a sense, that's the point. This scheme should be employed in its most robust form (i.e., with broad competitor definitions) when one reaches the conclusion that a positive economic dynamic is much more important than efficiency concerns.

But, actually, this scheme does have inherent limits upon costs. In practice, companies will need to make a product or deliver a service in order to be considered a competitor. This will limit how much they can spend on environmental improvements. They also must spend the money before they collect it, and some risk exists that their competitors will make environmental improvements in the interim that make collection of a premium impossible. This will provide some economic constraints in practice. But it will still leave opportunities for those confident that they can best their competitors in environmental performance without insane expenditures.

One can restrain this system (or other economic dynamic systems) to meet efficiency goals or other goals related to cost and still use it. The legislature could cap the dollars exchanged ex ante. This would, however, make the scheme much less dynamic and require government decision-making similar to that necessary to set a tax.

A more dynamic approach would be to evaluate the environmental expenditures and the environmental improvements realized ex post.[47] If monies exchanged became excessive, one might make a number of adjustments. One might decide that the competition had produced satisfactory environmental improvements for a particular pollutant and exempt that pollutant from the system in the future. One might determine that the costs generated were excessive and limit them going forward. Or one might conclude that insufficient improvement had come about and raise the premium paid to companies achieving superior pollution levels.

This ability to make ex post adjustments highlights a major economic dynamic feature of this system. Conventional regulatory approaches (including emissions trading) rely upon ex ante estimates of the cost a given set of requirements will create. These estimates are usually wrong. While data comparing pre-regulation cost estimates with actual post-regulation costs are not extensive, the data we have generally show great overestimation of costs.[48] Ex post judgments under an environmental competition statute would follow generation of accurate information.

Ex post facto judgment implies more adaptive efficiency. Recall that the adaptive efficiency concept required both an incentive to experiment (which the environmental competition statute provides) and a feedback mechanism to detect and correct failures. Ex post evaluation provides a feedback mechanism.

This ex post judgment mechanism's design would require the legislature creating the mechanism to grapple with the usual issues that bedevil environmental law. It would still have to decide whether to make a political judgment about what constitutes excessive cost or whether to delegate that task to an administrative agency according to some criterion. Delegation would require a political decision about what excessive cost meant. For example, does that mean cost disproportionate to benefit, or factories going out of business, or is no cost too much when pollution reductions save lives? Legislatures creating such a mechanism would still face the standard issues of environmental law. But no matter what criterion the legislature chose, subsequent judgments would involve more accurate information based on real world experience, not simply projections. And this system would avoid choking off innovation because of uncertainty and fear. It would invite innovation, and then check further innovation in a particular direction if it proved ill-advised.

This potential for ex post judgment also offers an avenue for public accountability, since the public can participate in the evaluation and decisions about modifications. If the legislature reviews the mechanism, the public can lobby the legislature. If administrative review takes place, then notice and comment rulemaking can include the public. This scheme, however, implies little public ex ante control over pollution levels. One might build in some ex ante democratic accountability by setting publicly determined pollution standards, using the environmental competition statute, to provide incentives to go beyond regulatory minimums. But absent ex post political judgments, direct public control over the magnitude of environmental improvement and expenditures would be lost.

This issue of public control will matter in any privatization debate. Privatization does imply some loss of democratic control in exchange for better public services.

Unorthodox legislation often attracts constitutional challenges. Laws requiring transfers of property from one individual to another have often attracted takings challenges—allegations that the law deprives a person of property without just compensation.[49] The majority of the Supreme Court, however, has stated that laws requiring monetary transfers without requiring transfer of particular property do not implicate the takings clause.[50]

Instead, the Court would decide whether such a measure offended substantive due process, a doctrine interpreting the constitutional prohibition against depriving an individual of property without due process of law. The Court typically upholds economic law having a "rational basis" against substantive due process attack. This law clearly has such a rational basis. It seeks to advance environmental protection by triggering competition to make environmental improvements.

Given the tradition of deference to economic legislation, it seems unlikely that a law that only demands payments from competitors with relatively large amounts of pollution would be deemed unconstitutional as fundamentally unfair.[51] The law would only operate prospectively, so it avoids the potential unfairness of retroactive legislation. Potential defendants can avoid the transfer payment by reducing pollution levels. Indeed, this system may set off a lot of environmental improvement without a lot of transfer payments if everyone scrambles to meet their competitors' levels.

One can design other systems seeking to privatize environmental law and/or to stimulate environmental competition.[52] And other ways exist to make environmental law more dynamic. The environmental competition statute demonstrates, however, that we can better maximize innovation if we focus upon economic dynamics rather than assume that efficiency enhancing reforms will facilitate more innovation.

An economic dynamic analysis invites systematic exploration of the idea of privatization. In particular, the contrast between government and private decision-making structures and how they respond to proposals for innovation points to the need to consider privatization. A central idea at the heart of the study of economic dynamics—recognition of the need to think through how economic incentives actually operate with some precision—aids our understanding of existing privatization and leads to fresh ideas for reform.

This chapter hopefully shows that thinking about privatization is useful and potentially productive. The coming chapters will explore other questions that economic dynamic analysis brings to the fore.

10
Making Public Environmental Decisions More Fair and Effective

Recall that the economic dynamic analysis in part II called attention to the plodding nature of government decision-making. It also emphasized how an economic dynamic—the ability of industry profiting from environmentally destructive activity to hire lawyers and consultants to oppose strict regulation—heavily influenced government decision-making processes. The analysis emphasized the peripheral role of environmental entrepreneurs.

That analysis necessarily implies that we need more serious thinking about how to make government decision-making less cumbersome and more dynamic. Of course, privatization circumvents this problem. But privatization will likely supplement, rather than supplant, governmental processes. Indeed, the most radical privatization proposal made (the environmental competition statute) involves significant and controversial changes and will therefore prove difficult to implement in the near term. And even this radical change might supplement, rather than supplant, regular government decision-making. Hence, economic dynamic analysis invites questions about how to make government processes more like private decision-making processes that tend to support innovation. Notice how radically different this question is from the question of how to make sure that government decisions efficiently use private sector resources. Exclusive emphasis upon this efficiency question tends to facilitate reform, such as more and more elaborate cost-benefit analysis, that can make government decision-making even more cumbersome than it is now. Whether or not one supports that sort of reform, economic dynamic analysis usefully calls attention to the need to at least consider how government processes might be less cumbersome and more fair and effective.

Toward Opportunistic Regulation

Currently, government agencies spend most of their time and energy dealing with people with a vested interest in opposing regulation—the owners of facilities causing the most serious pollution impacts. Government agencies spend relatively little time trying to figure out how to use regulation to benefit those who might have a vested interest in improved environmental quality—environmental entrepreneurs and companies interested in championing environmental innovation. Perhaps regulation should begin with a series of informal meetings with those hoping to profit from improved environmental protection. The idea would be to focus regulators on how regulation might introduce environmental innovation, especially positive radical innovation.

For example, suppose that EPA was concerned about electric utility emissions. Instead of meeting with the representatives of electric utilities prior to commencing rulemaking proceedings (as it does now), it would meet with purveyors of natural gas, solar power, and companies developing fuel cells. It would try to work with these companies to discover how it might use regulation to force utilities to make a choice between selecting some of these new technologies or losing out to new entrants.

Agencies must, however, eventually address the concerns of regulated industry and environmentalists. Still, beginning with detailed knowledge of what the possibilities are might strengthen EPA's hand in addressing both feasibility and cost concerns. It might also provide EPA with the vision necessary to provide environmental leadership in enacting legislation and otherwise formulating a long-term agenda. In evaluating the cost of environmental protection, EPA should consider not just the costs the regulated incur, but also the financial benefits that the purveyors of environmental innovation might realize. When EPA proposes to ban a substance, if a substitute is available that costs the same amount of money the net cost should be regarded as zero.

Taking full advantage of this approach will require political acceptance of the desirability of economically dynamic change. Current law focuses significant attention upon the interests of existing industry. While beginning with the innovators will help regulators avoid spurious arguments about technical infeasibility and exaggerated cost concerns (at least to some degree), radical technology forcing will prove difficult without clear statutory authority for it. A useful first step might be a statute directing agencies to consider whether regulations might create substantial new opportunities for new business as part of rulemaking.

Administrative Law Reform

Existing reform proposals focus primarily upon reducing judicial review's tendency to intimidate administrative agencies.[1] These reforms include the proposal of Professor Frank Cross of the University of Texas School of Law to abolish judicial review outright, and that of Professor Jerry Mashaw of Yale Law School to delay judicial review until after an agency rule is implemented.[2] These reforms may reduce the frequency of judicial reversal of agency decisions—a significant cause of ossification of rulemaking. This makes them very important economic dynamic reforms, worthy of serious consideration.

My primary aim here is to explain how these proposals remedy defects that the economic dynamic analysis of part II identified. These reform proposals rely upon an analysis that casts doubts upon the value of judicial review of agency action. A prime element of that analysis has been the tendency of judicial review to favor regulated parties at the expense of the public interest that environmental statutes often aim to serve. Professor Cross, in particular, emphasizes that judicial review favors those with the resources to litigate. In the environmental area, regulated industry files the overwhelming majority of challenges, notwithstanding the existence of some environmental groups that have lawyers on their staffs.[3] The economic dynamic analysis of environmental law adds to this account by explaining how resource exploitation tends to make agents of environmental destruction into powerful groups with superior resources over time.

I have already suggested that an economic dynamic analysis compares the relative rapidity of private decisions to deploy material innovation with the relatively moribund nature of public decision-making. The critics of judicial review find that it contributes greatly to this lethargy. Professor McGarity explains that in some cases (such as agency authority to eliminate hazardous substances under the Toxic Substances Control Act) judicial review has played a major role in stopping a regulatory program altogether. Professor Cross points out that regulated parties need not win a case in order to secure gains from judicial review. Even a loss causes delay, which benefits them, for delays in implementation occur during litigation and subsequent remand of rules. So prescriptions to eliminate or limit judicial review treat a major cause of administrative lethargy.

Part II noted that an economically dynamic system (like the free market) experiments even in the face of uncertainty, and tailors analysis to the needs of the decision-maker. Judicial review has made action in the face of uncertainty and tailored analysis more difficult. Judges have created the concept of hard look judicial review, which demands a "searching" review of an "administrative record" to make sure that the agency has engaged in reasoned decision-making. This demand has led to lengthy and difficult paper trials, designed not to meet the needs of agency decision-makers, but to provide a defense of decisions against unpredictable future judicial review. The need for this defense tends to make agency decision-makers quite cautious about writing rules that govern future conduct. Uncertainty about the future can make it difficult to construct a record that will satisfy judicial review, leading agencies to either forego rulemaking altogether or write anemic regulations.

Jerry Mashaw's recommendation to address this problem would enhance the adaptive efficiency of the regulatory system—its ability to experiment and then adjust to feedback. Professor Mashaw proposes to delay judicial review until after a decision directly regulating private parties is implemented. The current system discourages action by making judicial review into an abstract review of agency reasoning and response to voluminous input from regulatory parties. Mashaw's proposal would make actual experience with the program, not the court's assessment of the bureaucratic process producing the decision, the source of information about whether an agency acted arbitrarily. Hence, for example, in determining whether a mandate creates a feasible requirement, a court would review actual corporate attempts to adapt to the requirements, instead of the quality of the agency's written response to corporate claims that requirements were not feasible. While not conceived in this way, Mashaw's proposal makes an administrative process demanding innovation resemble in some ways the private sector process for experimenting with innovation.

A full evaluation of arguments to limit or eliminate judicial review lies beyond the scope of this book, but a brief explanation of why many administrative law scholars doubt the value of judicial review seems necessary. Judicial review aims to serve sufficiently important values to raise questions about any proposal to limit it in order to facilitate effective implementation of public programs. Judicial review seeks to ensure administrative fidelity to the law and non-arbitrary choices in the exercise of discretion. My claim that these proposals merit serious consideration rests in part on scholarship raising serious questions about whether judicial review in fact serves these values.[4]

Frank Cross argues that judicial review does not serve the rule of law. Often statutes are vague enough, he claims, that judgments reversing agency statutory interpretation simply substitute one contestable interpretation for another, rather than correcting a clear legal error. Furthermore, legal scholars and social scientists have found that judicial statutory choices often correlate with the ideology of the judges.[5] Some reason exists to question the claim that judicial review promotes rule of law values.[6]

Even more scholars doubt the value of judicial review under the arbitrary and capricious standard. The Administrative Procedure Act authorizes such review to correct arbitrary agency action. But the courts have developed a hard look doctrine that many scholars find leads to arbitrary judicial decision-making. And some scholars believe that bureaucratic and democratic checks will deter truly arbitrary decisions anyway, even though all decisions in a controversial area will strike some observers as unreasonable.

Proposals to abolish or greatly limit judicial review would improve the potential of agency decision-making to keep up with growing environmental threats. Scholars have made sufficient arguments against the value of judicial review to make these proposals worthy of much more widespread consideration than they have received to date.

Focusing upon slovenly bureaucracy as a problem should lead to exploration of other avenues of streamlining administrative processes as well. For example, one might consider changing the fundamentals of administrative procedure itself, not just judicial review. Currently, the administrative law system reflects a devotion to the principle of open and unlimited access to the agency. This approach has much to recommend it. Openness invites participation. It certainly stimulates many hours of discussion and volumes of written comments. But this input has the potential to overwhelm the agencies, greatly slowing the decision-making process.

Scholars should think about the value of limits upon the volume of comments that interested parties may submit and the amount of time agency staff can spend meeting with people. If agencies need to adequately address problems stemming from diffuse pollution sources, perhaps the agencies or Congress should limit the amount of resources an agency may devote to responding to public input in any one rulemaking proceeding.

Agencies cannot limit their use of resources very readily now. First, agencies have a legal duty to respond to all significant comments. So, bureaucrats must read all comments and respond, in writing, to most of them. This remains true whether the comments have great or little value to agency decisions. Second, refusing to meet with somebody with a legitimate interest in a regulatory proceeding may cause great offense and corresponding political problems. Indeed, since EPA faces a risk of judicial reversal if anybody challenges a final rule, its staff often meets with parties over and over again, in hopes of making enough compromises to convince potential challengers not to litigate. This deluge of meetings occurs even when the agency does not employ a formal regulatory negotiation to resolve rule-making issues.

Limiting the volume of communication will not interfere with sound, deliberative decision-making. Consider the possibility of limiting written comments. Top political appointees, those nominated by the president and confirmed by the Senate to make tough decisions, cannot read all of the comments submitted now in many regulatory proceedings. EPA staff members review submitted comments, but even the most dedicated staff would find it difficult to fully absorb and respond to all of the comments EPA currently receives.

Federal courts generally limit the length of briefs (written arguments) that lawyers may file. While lawyers may chafe under these restrictions, the page limits force litigants to focus on the most important issues. The limits also enhance the judicial decision-making process by decreasing the likelihood of a decision resting on some misunderstanding of a long and complicated brief, or the failure of a party to address some erroneous argument buried in her opponent's monumental document. It also makes it easier for the judge to identify and think about the most important issues, because parties omit less important matters from their briefs.

EPA staff has generally become quite used to inviting unlimited public participation. I would expect most experienced staffers to oppose any real change in this tradition, which makes a lot of sense given the bounded rationality of agency employees. Many government employees would rather work very hard to wade through all possible comments than miss a problem in a rule that they might have anticipated. This merely highlights, however, the considerable difference between public and private decision-makers who might have some capacity to induce innovation. Government employees, since they work in a static environment that does not substantially penalize inaction, have become accustomed to a pace of operation that would embarrass and perhaps bankrupt any business facing fast moving competition. By contrast, making decisions that do not work out (as defined by a quite capricious political environment) can significantly affect the agency's budget and the future of its employees. Hence, EPA (like most government agencies) is much more risk-averse than many private firms. My analysis of the long-term problems of environmental decision-making keeping up with private, environmentally destructive decision-making suggests, however, that the understandable views of very intelligent staffers may prove dysfunctional for the physical environment. Moreover, reducing the volume of comments submitted will not significantly increase the probability of EPA missing something important. No matter how many comments EPA considers, some unanticipated problems will arise. Conversely, very often the key information to avoid significant anticipated problems can be conveyed in less space.

Structuring limits, however, poses a challenge. Rulemaking often involves more complicated detailed decisions than judicial decision-making. EPA must not only determine if a proposed rule is legal and non-arbitrary (the court's role), but also whether it is the most desirable of the legally available alternatives from a law and policy standpoint. The limits must allow parties to address the most important issues, without being so long as to invite objections to every detail of a proposed rule. The Federal Rules of Appellate Procedure meet this structural challenge by imposing uniform limits while allowing courts to lengthen the limits in an appropriate case. A similar approach could apply to rulemaking.

The open access principle also conflicts with the principle of equality in participation, because of the economic dynamic. Generally, the small number of professionals defending the environmental interests in rule-making cannot meet as often with agency staff or file as much commentary as the regulated industry. Environmental entrepreneurs also cannot compete with the regulated industry in terms of the sheer volume of written submissions and face-to-face contact with decision-makers.

Scholars could, however, think about systems based on equal participation rather than equal access. Instead of allowing agencies to passively respond to the input they get, the government would regulate the input to produce roughly equal representation of the principal interests at stake.

A modest step in this direction might involve making participation easier for affected citizens. The economic dynamic analysis points out that regulated companies send employees and paid consultants to meetings and hearings. These gatherings occur during business days. EPA holds most of its meetings in Washington, D.C. or a handful of other locations where it has offices. EPA usually notifies interested parties of these meetings through Federal Register Notices.

None of this poses a substantial barrier to participation by national environmental groups or industry. Both have full time employees to attend meetings in Washington, D.C. and monitor the Federal Register. But often the people most directly concerned about the results of EPA regulatory proceedings live near a plant scheduled for regulation, far away from Washington, D.C. They may not have lawyers and may not regularly peruse the Federal Register. They may not have money to fly to Washington, D.C. And they may work during the day and often cannot take time off from work to participate in regulatory proceedings.[7]

EPA has, to its credit, occasionally made special efforts to involve local communities in its regulatory program. And it has begun to think about doing more. More could indeed be done. In the early 1990s, EPA began writing a rule addressing toxic emissions from the chemical manufacturing industry. New Jersey, Houston, and the Baton Rouge-New Orleans area contained most of the regulated plants. Upon the request of NRDC and local environmental groups in Louisiana, the agency held public hearings on the rule in Louisiana.[8] Hundreds of people who breathed in chemical plant emissions daily, many of them people of color, attended. The hearings took place in the evening, when people who work for a living doing things having little to do with environmental policy could attend.

These hearings, however, were exceptional. While EPA has done something like this on other occasions, EPA usually conducts regulatory proceedings far from affected communities. EPA genuinely welcomes public participation, but it lacks the resources to hold frequent hearings in the communities its rules affect. More resources could make local hearings a more frequent part of regulation. Generally, funding for public involvement has declined over the years.[9]

EPA might be able to do more to let the public know about regulatory proceedings. Even those who cannot attend a meeting might participate by sending in comments, or by talking with agency officials on the phone, through video conferencing, or through email. EPA might systematically create databases pinpointing the predominant locations of plants they plan to regulate and develop lists of local contacts to help them locate and advise community leaders about pending proceedings of interest to them. EPA has already begun posting significant amounts of regulatory materials on its Web site, thus improving public access, at least for those who own or can get to a computer.

While greater involvement of local communities may change the economic dynamics somewhat, even participating local communities will find effective participation difficult without access to technical help. Hence, some means of funding local participation remains vital.

A thorough commitment to equality of participation would require more radical reform. To make an understanding of what full equality might entail, it will help to sketch a proposal, which I will call the Equalization Act. While the principles of equalizing and limiting the volume of input do not necessarily require the acceptance of this particular proposal, it will make the idea of equal participation more concrete.

Congress could limit regulated parties as a group to no more than five hundred pages of comments in any one rulemaking and impose the same limits upon environmentalists and environmental entrepreneurs. It could limit its staff to forty hours of meeting time with each group for each rulemaking. This proposal combines equal participation with limited input to ensure efficiency.

Congress could also enact limitations to ensure equal access to the expertise necessary for effective advocacy before agencies and reviewing courts. It could require that each group provide a budget for their expected expenses for the rulemaking, including potential subsequent judicial review. This budget would include the hours of staff time devoted to the rule, the salaries and benefits that each relevant staff person gets, the cost of outside lawyering and scientific consultants, and the cost of travel and other services needed. This budget would then become an expenditure limit. Exceeding the cap would result in a loss of the right to seek judicial review and a monetary fine. Furthermore, the party with the largest budget (most likely the regulated industries) would have to pay a tax equal to the difference between their budget and that of other parties (most likely environmental groups and environmental entrepreneurs). This tax would be transferred to the environmental group and entrepreneurs to pay for hiring of additional experts.

This Equalization Act would assure that decisions reflect the results of more or less equal participation. It would limit the chances of rulemaking simply reflecting a response to overwhelming force.

The monetary transfer provisions of this Equalization Act raise serious free speech issues under the First Amendment to the United States Constitution. This proposal has value anyway for several reasons. First, it highlights a key issue that an economic dynamic theory brings into focus, the problem of those who profit from environmental destruction thereby acquiring disproportionate influence upon the law of environmental protection. Second, the monetary transfer provisions might pass muster under existing precedent. Finally, free speech law may evolve to make something like this possible, if it does not meet the strictures of current law. The free speech cases that this proposal implicates merit rethinking in light of economic dynamics.

The monetary transfer provisions do not directly limit industry speech. Industry remains free to propose an unlimited budget under this proposal, so it may hire as many experts as it wants to assure high quality input. Limiting the amount industry could spend directly might conflict with *Buckley v. Valeo,* which struck down limits on campaign expenditures under the First Amendment.[10] *Buckley* treated limitations on the expenditure of money used in communicating political messages as tantamount to limiting speech itself. The Court would probably treat direct mandatory limitations of lobbying expenditures as a limitation on speech as well, and subject such limitations to the strict scrutiny that doomed the campaign expenditure limitations. Hence, the lack of direct limits upon expenditures in the Equalization Act may prove essential to its constitutionality.

Nevertheless, this proposal would likely face a First Amendment objection, based on a claim that it may force industry to finance its opponents' speech. The Supreme Court has held that free speech principles forbid government from compelling a person to finance an opponent's message. In the leading case, *Abood v. Detroit Board of Education*,[11] the Court held that a union of public employees may not compel dissident non-members to finance "ideological union expenditures," because this involves financing expression of opinion with which the dissidents disagree.

The transfer provisions, however, do not necessarily compel financing of environmentalist speech. Industry remains free to spend as little as the environmental group does in order to escape the transfer obligation. The Supreme Court has held that absent compulsion, the Constitution does not bar spending other people's money in ways that the contributor would disapprove of. The Court held, for example, that corporate speech does not violate shareholders' rights, because dissident shareholders do not have to buy shares.[12] Similarly here, dissident companies do not have to outspend environmental groups on lobbying. So they remain free to avoid financing environmentalist speech by limiting their own spending to an amount equaling that of their opponents.

The line between unconstitutional compulsion and constitutional choice may not be sharp, however. The Court might evaluate the Equalization Act in light of its statement in *Buckley* that the First Amendment does not countenance the restriction of some voices in order to "enhance the relative voice of others." This should not literally bar the transfer provisions, since the provisions do not limit corporate expenditures on lobbying. But the spirit of this statement may influence how the Court views the provision, which obviously has equalization of resources devoted to rulemaking as a goal. The Court may be troubled by the notion that a company must limit its own lobbying expenditures, and thus its speech (in the Court's view), if it wishes to avoid financing opponents' speech.

Buckley justified its rejection of an equality goal for speech regulation by pointing out that the First Amendment was "designed to secure the widest possible dissemination of information from diverse and antagonistic sources."[13] But a system of free access to administrative proceedings does not provide for the widest possible dissemination of information from diverse sources. Development of information and arguments that influence regulatory proceedings requires money to pay professionals to develop legally and scientifically sophisticated arguments. In many cases, only the national environmental groups and regulated industries have the necessary resources. And even the national environmental groups lack the resources to monitor every important regulatory proceeding, leaving some cases where only the regulated industry can make its voice heard. Local neighborhood groups with a vital interest in the emissions from a factory in the neighborhood often have no access to scientists or lawyers to advocate on their behalf, so their views are rarely developed in detail and disseminated.

Congress could write transfer provisions designed to stimulate more thorough consideration of innovation's potential. The Equalization Act might allocate a transfer payment to competitors or suppliers for the regulated industry who might want to argue for stricter limits to create markets for environmental innovation, over and above that required to help fund environmental groups. Their views, when they come forward at all, often receive short shrift, because many companies with new technology cannot devote resources to lobbying comparable to those that already established, regulated companies possess. This, however, involves some tension with the equalization principle, since environmental groups and environmental entrepreneurs may share the same interest.

This proposal would raise a host of issues. For example, one would have to develop a legal test to make sure that only genuine environmental groups received funding destined to assure equal representation of environmental interests. Industry might otherwise establish "environmental groups" to take weak positions in regulatory proceedings with funds devoted to environmental protection.

This funding of environmental groups may seem objectionable to those who view environmental groups as just another special interest group with too much influence already. In light of the broad support for stringent environmental protection among the public, this objection is questionable. Moreover, the aim here is not to secure representation of every single individual in the country's views or to have groups divining some general public interest. Rather, the agency generally applies legal criteria chosen by an elected Congress to a concrete problem to come up with specific standards.

It needs diverse views about the appropriate standards in that context. The regulated industry will surely adequately represent the views of any individuals who share its preference for lax standards. As long as contrary views have been adequately represented, the agency will have a good understanding of the issues and must assume the responsibility for implementing the public interest as defined by the implementing legislation.

Also, questions would arise about the participation of sub-groups. On the industry side, different companies may not always share the exact same views about rules. The same might be true of different environmental groups and different environmental entrepreneurs.

The companies would be in a good position to work out subgroup participation's place in the industry budget themselves. After all, each company can propose an unlimited budget. The companies would just have to pony up more to environmental groups if they do this. This may encourage some coordination to reduce costs, but companies with really distinct interests from the majority may keep their budgets intact.

The real problem would come in distributing money to environmental groups and perhaps to environmental entrepreneurs. Often, environmentalists and entrepreneurs can agree on how to divide up funding. One might need a set of legal rules to address those occasions when this fails to occur. One might divide up money going to environmental groups on the basis of membership numbers (indicating degree of public support). Other arrangements would also be possible.

Hopefully, readers can develop other proposals for limiting input in a way that enhances equality of representation without stifling adequate debate about regulatory proposals. My point is not necessarily that this Equalization Act provides the only good answer to the difficult problems such a task presents. Rather, the point is that an economic dynamic analysis leads one to inquire whether more might be done to free agencies to make decisions and to equalize the pressures upon them. Currently, almost all of the scholarly discussion and movement in administrative law leans toward more jawboning, more talk, more analysis, more consideration, and more comments.

From an economic dynamic perspective, one contrasts the decision-making structure of agencies to that of firms and investors pursuing innovations to improve our material well-being. That comparison highlights the enormous pressures from parties not interested in innovation and the consequent slovenly pace of agency decision-making. This should lead policy analysts to ask about streamlining and improving agency capacity to make timely and frequent decisions, whether or not they like the proposals presented here to stimulate thinking about this problem.

Alert readers might ask whether the appropriate reform of administrative decision-making should simply remove all procedural constraints on the agency. After all, private parties considering adoption of material innovations operate without significant legal mandates about decision-making processes. The logical market emulation move might be complete removal of constraints upon agency decision-making.

This is well worth thinking about. But the requirement that the agency consider comments from all affected parties affords regulated parties and regulatory beneficiaries a degree of dignity by requiring that agencies consider their input. Since the agencies make decisions that coerce regulated industries and should benefit others, this seems necessary. Also, a reasonable flow of good information and ideas should improve the regulatory process (even if an ocean of input provides little value). But the main point is that an economic dynamic analysis invites more attention to this kind of question. Serious widespread debate of how to streamline agency procedure and equalize participation would involve a substantial change in the current regulatory reform debate.

Bypassing Administrative Decision-Making

The environmental legislation of the early 1970s relied upon administrative decision-making for almost every specific decision. Happily, this has begun to change, to a degree. The 1990 Clean Air Act Amendments, for example, contain many more specific Congressional decisions than previous environmental legislation. This left fewer decisions for state and federal agencies to make, although the statute still left them with the lion's share of decision-making responsibility.

Congress, for example, set numerical limits for sulfur dioxide emissions for each unit of major power plants (but authorized emissions trading to increase flexibility), established a presumptive schedule for phaseout of ozone depleting chemicals, and set specific numerical standards for automobile emissions. While the statute still requires rulemaking to fill in some of the details and adjust to future problems and opportunities, the existence of such a detailed set of decisions has made the rulemaking for these programs unusually expeditious and successful.[14]

The difficulties Congress experiences in making large numbers of detailed decisions will limit the extent to which detailed congressional decision-making can lessen the need for administrative decision-making. Nevertheless, efforts to get Congress to make as many detailed decisions as possible help improve the economic dynamics of environmental lawmaking.

Other bypassing mechanisms might be possible. For example, Congress could require that industry achieve the lowest emission level feasible while barring EPA from writing guidance or rules that explain precisely what this means. Industry would then have to make their own decisions about what is feasible. If a government or citizen enforcer could show that they failed to use a feasible technique that would achieve a lower limit, they would pay a heavy fine, so they would have an incentive to be fairly aggressive in implementing controls and forced to make continuous improvement. This would create a pressure similar to the constant pressure to improve that a competitive market creates.

Of course, this creates great uncertainty for industry. Industry likes rulemaking because it does not want to guess about what it must do. But from an economic dynamic perspective this whole effort to avoid uncertainty seems misplaced. Industry always faces uncertainty in making its products and services. It does not know exactly what the public wants or exactly what its competitors might do. As a result, it may have to innovate and work very hard to improve in order to be successful under uncertain conditions. Perhaps environmental law should operate from a similar economic dynamic premise. The general legal standard should create the equivalent of consumer demand upon the industry for environmental improvement. Industry should then have to guess about how to meet that demand, with significant economic consequences flowing from wrong guesses. Again, much can be said against this proposal. But economic dynamic analysis, in particular the contrast between free market and government decision-making processes, suggests that we have become too complacent in accepting plodding administrative decision-making and need to think more creatively about possible alternatives. Recognizing that a variety of processes might bypass administrative agencies either partially or altogether will stimulate more creative thought about how to improve economic dynamics.

International Law

Economic dynamic reform might also improve international environmental law. Economic dynamic analysis focuses attention on change over time, and new environmental problems arise over time. In the 1970s, the world focused a lot of attention on endangered species. While this problem remains with us, ozone depletion provided a fresh and urgent issue in the 1980s. By the 1990s, climate change loomed large on the international agenda. Recently, fears about reproductive toxicity issues have led to some international regulation of persistent organic pollutants.[15] Each new issue now requires a fresh international agreement by consensus. And this consensus, if it emerges, demands at best a long sequence of domestic measures to regulate private parties contributing to environmental problems.[16]

Perhaps the time has come to develop some general principles of international environmental law and institutional mechanisms to force environmental innovation more directly, apart from any problem specific treaty. National governments, for example, could agree to a treaty requiring all private companies to employ innovations effectively addressing international environmental problems, if they have the capacity to do so. The treaty might establish an international court to hear claims that a company has failed to employ an available innovation to address an international environmental problem. Such a procedure would only prove effective if the court's rulings could be enforced against regulated companies. The treaty could provide that any company that failed to employ an innovation that the court found it should employ would face a trade ban on its goods.

While the treaty would specify that any problem addressed in an existing international treaty constitutes an international environmental problem, it would leave the court free to decide what new problems are genuine international environmental problems in the future. This would greatly change the economic dynamics of regulation, because it would eliminate doing nothing as a default option. Currently, when countries fail to agree to a treaty effectively addressing an international environmental problem, private companies may do nothing to address it. Indeed, the existence of a do-nothing default option gives private companies an incentive to discourage their governments from entering into treaties threatening their economic interests. With the existence of some general principles that apply not to nations, but to individual companies, the default option will seem much less attractive to companies contributing to international environmental problems.

Indeed, companies might prefer an effective international treaty to complete reliance upon the court procedure. Agreement to an international treaty may create a national implementation strategy that seeks progress on the environmental issue it addresses. An international court might consider this strategy in deciding whether a technological innovation is effective and available. So, a treaty could provide some sort of a defense for a company brought before this court. And, as a practical matter, plaintiffs would bring more claims to this international court when the international treaty-making process was not producing results than when treaty making was producing real progress.

This idea raises some economic dynamic issues. One could give companies a greater interest in supporting treaties and their implementation by creating a formal defense for companies in compliance with applicable requirements implementing a problem specific international treaty. This defense could provide that if a treaty existed addressing the pollutant giving rise to a claim, a national program addressed this pollutant, and the national program effectively regulated the defending polluter, the court would dismiss the claim. This would involve an economic dynamic trade-off, however. Absent dismissal of the claim, the court might impose a more stringent obligation on a company than the national plan would. The creation of the defense would lessen corporate incentives to innovate, but create better incentives for them to favor creation of international treaty regimes.

This idea of direct international regulation of private parties will appear unorthodox to international legal scholars and troublesome to diplomats. Traditionally, international law addresses relationships between states rather than regulating private parties. Countries may feel reluctant to allow an international tribunal to adjudicate the rights of private parties.

Precedent exists for direct international regulation of individuals under the law of human rights. While human rights treaties and customary law impose duties upon states, these duties have resulted, at least since the Nuremberg trial of Nazi war criminals, in the imposition of liability upon individuals.[17] Indeed, some of the modern human rights treaties outlaw individual actions harming human rights.[18] The United States, while sometimes slow to ratify human rights treaties, has pioneered judicial remedies against individual violators of human rights. In *Filartiga v. Pena-Irala*,[19] and a number of other cases,[20] United States courts have allowed torture victims to sue the individuals who tortured them under the Alien Tort Claims Act.[21, 22]

Nevertheless, countries may view direct international regulation of private companies as a threat to their sovereignty. They may not like the idea of an international tribunal determining what environmental measures their companies should take. In principle, however, this arrangement does not involve a significantly greater loss of sovereignty than that provided for in other international agreements. For example, when developed countries agreed to phase out ozone depleting chemicals, they could no longer, consistent with the Montreal Protocol, leave facilities producing the banned chemicals for the domestic market open. They yielded sovereignty to that extent. Countries can likewise allow some international control over their producers based on some general principle. The real debate should focus on what principle is appropriate.

In this connection, states already have a duty under customary international law to prevent transboundary pollution.[23] In fact, however, private companies and individuals, not states, often cause the pollution. It would be quite logical to enforce compliance with this underlying norm directly against violators.

Indeed, in a leading case establishing the duty to prevent transboundary harm, the *Trail Smelter* arbitration,[24] the arbitrators ordered that the Trail Smelter, a privately owned facility, "be required to refrain from causing any damage" and "shall be subject to some control" of emissions. This establishes, at a minimum, that states may agree to create arbitration mechanisms that produce orders running against private companies.

Other possibilities exist beyond creating international jurisdiction and law to require private parties to refrain from pollution with transboundary effects. Following *Filartiga*, countries may wish to create national laws creating causes of action against companies that cause transboundary harms. Indeed, some commentators have argued that some types of environmental harm constitute human rights violations and are actionable under *Filartiga*.[25] And some transnational litigation regarding environmental harms has already taken place, although it has often proved unsuccessful.[26] Transnational environmental law could expand through development of a substantive standard reaching most failures to prevent transboundary pollution, when an effective preventative measure is available.[27]

Economic dynamic analysis does not require adoption of the particular solutions that I have outlined. But it does require serious thinking about the questions these proposals seek to answer. If administrative law and international consensus on each individual environmental improvement provide inadequate continuous incentives for innovation and compare poorly with the decision-making structures governing employment of innovations improving material well-being, one must think seriously about how to reform these processes. And the reforms adopted should, under this theory, improve the economic dynamics, providing for more frequent and effective decision-making.

This involves a major rethinking of regulatory reform. Efficiency-based regulatory reform evinces little concern with the effectiveness and frequency of environmental decision-making. It views each regulatory decision as a kind of transaction and asks how to make sure that each transaction efficiently uses private sector resources that the regulation will affect. It cares not one jot about whether the system as a whole produces lots of decisions or very few. By contrast, economic dynamic theory takes the free market model of frequent, decentralized decision-making seriously, as a source of energy for a dynamic system, even in the face of some inevitable bad decisions. It demands serious answers to the question of how one can make the environmental decision-making system as a whole perform with reasonable vigor.

Improving Regulatory Design to Stimulate Change

Previous chapters focused upon two types of process questions—questions about privatization of environmental decision-making and streamlining and improving the fairness of overly awkward public decision-making processes. When agencies write regulations, either traditional regulations or the rules governing emissions trading programs, they make important decisions about regulatory design that influence innovation. Hence, another question that concern for innovation invites arises. How can regulatory design encourage more innovation?

Analyzing Traditional Regulation's Dynamic More Carefully

Efficiency-based regulatory reform aims at grand generalizations about traditional regulation's failings and the contrasting virtues of "economic incentives." These generalizations include the claim that traditional regulation discourages innovation. While traditional regulation has significant failings in this regard, analysis should not rest there. Analysis of regulation's economic dynamics, what precise incentives they provide and how they interact with free market incentives to influence firms operating under bounded rationality, can clarify the nature of traditional regulation's dynamic failings, and account for the regulations that have, in fact, spurred significant innovation. This analysis leads to a firmer grasp on how regulatory design might improve the economic dynamics of environmental law.

Many studies describe traditional regulation as posing a "barrier" to innovation. The barrier imagery suggests that once this "barrier" is removed, environmental innovation will flow. These studies, however, rarely explain what they mean by a regulatory barrier, so the claim remains puzzling. And the conclusion that traditional regulation poses a regulatory barrier provides little guidance about how to improve regulatory design.

Careful attention to the economic dynamics of regulation helps create a conceptual framework that can aid analysis of traditional regulation. In particular, economic dynamic analysis can clarify the concept of regulatory barriers and lead researchers to more careful analysis.

An earlier chapter explained that the free market does not generally provide incentives for environmental innovations that cost money to implement. This would suggest that the description of regulation as posing barriers to innovation will usually prove misleading. If no environmental innovation would occur without a regulation, then regulation cannot properly be described as providing a barrier to environmental innovation. Rather, regulation usually provides an incentive to make an environmental improvement that would be wholly lacking absent the regulation.

A portion of a recent Environmental Law Institute (ELI) study provides an excellent illustration of this.[1] The ELI researchers found that dry cleaning establishments often lacked the financial capability or the knowledge necessary to employ environmental innovations that would greatly reduce air emissions from their operations. For the most part, this industry consists of relatively small businesses with little money available to hire environmental experts. The study shows that absent regulation the industry would not innovate, so the regulation cannot prevent innovation that would occur without it. This implies that the regulation poses no barrier to innovation. Indeed, the study shows that regulation has stimulated innovation in pockets of the industry.

The regulation of dry cleaning, however, did not stimulate innovation as effectively as it could have, because it contained requirements that fairly standard technology could satisfy. This would suggest not that the regulation posed a barrier to innovation, but that it failed to stimulate widespread innovation. Indeed, the study, at one point, drops misleading barrier terminology found in the conclusion and introduction and properly describes the issue as how to improve regulatory design to make it perform better as a stimulator of innovation.[2]

Researchers regularly use the "barrier" terminology when their data support the milder inference that the studied regulation fails to stimulate environmental innovation as effectively as it could. This particular error, confusing a failure to provide the best possible incentives for innovation with the posing of an actual barrier to innovation, however, has become pervasive in the field because of a lack of clear theoretical thinking about economic dynamics.

This distinction between a barrier to innovation and a less than perfect stimulator of innovation matters in debates about policy. Characterizing regulation as a barrier to innovation suggests that getting rid of the regulation might improve things. Perhaps more importantly, incorrect claims about barriers facilitate arguments that almost any change will lead to greater innovation. Characterizing a regulation as not providing the best possible incentives for innovation leads to more careful inquiry as to how to improve regulation.

Once analysts realize that almost all regulations described as posing barriers to innovation actually only involve inadequate stimulus for innovation, several questions present themselves. First, do regulations ever pose actual barriers to innovation? Second, in cases where regulation poses no barrier, but fails to stimulate innovation, why does it fail? Third, how can one design regulations that stimulate more innovation? The misdescription of so many regulations as posing barriers to innovation truncates discussion of these questions. It leads to environmental policy based on sloppy free market imagery, rather than careful analysis of economic dynamics.

A regulation can, in theory, pose a barrier to innovation only when an innovation would take place in the absence of the regulation and the regulation prevents the innovation. The economic dynamic analysis helps predict when the free market will tend to provide incentives to produce an environmental innovation. The free market may provide an incentive for environmental innovation when employing environmental innovation actually saves firms enough money to make the investment worthwhile from a purely financial point of view.

If an analyst can identify a set of profitable measures that firms would employ absent regulation, the analyst might inquire into whether relevant regulations prevent seizing these opportunities. Researchers must exercise great care in evaluating claims that regulations pose barriers to profitable environmental innovations, however. Regulated companies' lobbyists frequently claim that regulations pose barriers to innovation, because this aids their companies' general effort to reduce expenditures to comply with regulation.[3] But analysis of such a claim must involve determining whether the regulation actually prevents using a profitable innovation, a really new technique. If no new technique is involved, the regulation does not discourage innovation, but simply discourages use of a particular old technology.

Regulators should also evaluate whether the regulation prohibits use of an innovation that actually improves environmental performance, i.e., performs better than the technology upon which regulation is based. Otherwise, analysts will tend to characterize regulations as preventing environmental innovation when they prevent use of environmentally inferior technology. Companies innovate all the time in ways that harm the environment. The types of innovation that the regulatory system needs to encourage, though, are those that improve, not diminish, environmental protection. Claims that regulation blocks environmentally superior innovation should be the focus of inquiry. Analysts should distinguish between industry preferences for less stringent regulation for the sake of lower cost and claims that regulation has impeded the use or development of environmentally superior technology.[4]

Researchers must take care to distinguish cases where regulations demand something that may supplement an innovation from cases when they somehow prevent employment of the innovation. For example, suppose that a company finds that a pollution prevention project would reduce emissions at one of its sources by 15 percent. EPA requires a 98 percent reduction from this source. EPA has identified an end-of-the-pipe technology capable of realizing the 98 percent improvement. The company may claim that EPA's regulation requiring the 98 percent reduction poses a barrier to innovation. But a requirement for a 98 percent reduction does not legally preclude employment of the pollution prevention measure. The company may combine the pollution prevention with the end-of-the-pipe control to produce more than a 98 percent reduction, which the regulation allows. This should cost less money than making the minimum 98 percent reduction, because the pollution prevention measures save the company money (otherwise, the company would not employ the pollution prevention measure even absent the regulation). Indeed, the company may find itself able to use a less expensive control technology than EPA envisioned and still meet the 98 percent reduction requirement, because the pollution prevention measure provides for some of the required reduction.

Researchers must account for the possibility of combining an innovation with the expected compliance measure, as well as the possibility of some substitution or modification of the method of compliance.

If an EPA requirement poses a barrier to innovation, researchers should analyze how it poses a barrier and why it poses a barrier. For example, if a 98 percent reduction requirement did somehow prevent employment of the pollution prevention innovation, it would be important to consider whether this rigidity flowed from the decision to seek more reduction than the innovation would produce or from some kind of need for monitoring.

In short, a thoughtful, revealing analysis of real regulatory barriers (if they exist) would include an explanation of why the researcher thought a company would employ an innovation absent regulation and how and why the regulation prevented employment of the innovation. This would require a subtle appreciation of both the economic dynamics involved and careful analysis of the law and the facts. This kind of analysis would help enrich discussion of how to improve regulation's economic dynamics.

The New Source Review Debate

The Clean Air Act generally imposes stricter requirements upon new air pollution sources than it imposes upon existing sources. Congress wrote this difference into the Act, because retrofitting existing sources to control pollution costs much more than designing control into new sources. Congress expected that over the years, cleaner new sources would replace dirtier existing sources.

Scholars have criticized the Act's treatment of new sources as discouraging innovation. They point out that strict requirements for new sources provide a disincentive to replace older equipment with new (and presumably cleaner) equipment.[5] This would imply that new sources might not replace existing sources over time and that the congressional policy needs rethinking.

This analysis views new source requirements as a regulatory barrier to replacement of old sources. This analysis suffers from the same inadequacy identified earlier. The analysis assumes that new sources would replace old sources, but for the regulations. This assumption has for the most part cut off serious study of the question of whether regulatory requirements have greatly influenced corporate decisions about whether to build new plants. We actually know very little about this.[6] Nevertheless, state and federal environmental protection agency officials often assume that the new source requirements have slowed replacement of old, dirty facilities. And we know that in some industries, the electric power industry being the most prominent example, old plants have been kept running an awfully long time.[7] But absent careful study of whether the economics and law invite new source replacement absent new source requirements in the CAA, we have no way of knowing whether the CAA has really discouraged modernization. Careful study of how corporate managers have arrived at decisions to continue to run existing facilities or to build new ones would help.

We do know, from numerous studies, that environmental control costs usually play only a minor role in decisions about where to site facilities.[8] This would surprise many state regulators who often fight to relax state standards to attract new industry and state legislators who pass laws forbidding state pollution control agencies to implement standards more stringent than federal minimums.[9] But environmental control costs are not high enough to matter much to most corporations building plants. Corporate managers generally care much more about proximity to markets, available skilled labor, and access to transportation for their goods. This might lead one to expect that CAA requirements may have a negligible impact upon decisions about whether to replace existing facilities with new ones. But a final conclusion must await the serious study that careful thinking about the economic dynamics of regulatory barriers invites.

Economic dynamic analysis also aids thinking about how to handle the problem of regulatory design to respond to potential barriers to innovation. Economic dynamic analysis, as suggested in the introduction, involves considering how both free market incentives and regulatory incentives influence managers of firms, who presumably use bounded rationality to make decisions. It requires some precision about this, and eschews considering either factor alone. Analysis should begin with the question of how private market incentives might encourage plant replacement absent regulation. This technique helps clarify the issues at stake in new source review. Private markets might encourage replacement of older equipment absent regulation, because old equipment wears out eventually and needs replacement.

Furthermore, newer equipment sometimes offers more production at less cost, so at some point, it pays to modernize.

Suppose that the CAA only applied new source standards to brand new factories. How might a corporation eager to have new equipment respond? The corporation might simply replace old equipment with new equipment at the old plant, or it might simply rebuild the old equipment with new parts.

This helps explain why new source review requirements apply not only to new facilities, but also to "modified" facilities under the CAA. From an economic dynamic perspective, strict regulation of renovation must apply in order for new source review provisions to succeed in obtaining upgrades in pollution control as facilities are rebuilt or modernized. Otherwise, owners of older facilities can replace worn-out equipment without delivering the air quality benefits new facilities can bring.

If modification provisions capture all renovations and replacements, however, then owners of older facilities face a very different choice than that involving strict controls only of brand new plants. They must either put up with equipment that is wearing out (which becomes impossible at some point) or they must comply with the stricter requirements. Since production with no renovation may be impossible, such a strict policy might encourage building of new facilities. Notice that this strict approach would make patching up old facilities without new source review impossible.

The modification provisions of the CAA, however, do not capture all renovations. The Act defines modifications to only include projects increasing emissions. And EPA, by regulation, has exempted replacement of old equipment in at least some cases. This means that renovation projects that do not increase emissions, or replace old, worn-out equipment with new equipment of the same type, do not always trigger new source requirements. So some modernization projects and a lot of patching up of old equipment escape new source standards.[10]

One can ask about whether this makes sense from an economic dynamic perspective. From a convenience perspective, it makes perfect sense. The exemptions for some renovations allow owners of existing facilities to continue to operate old plants and patch them up without becoming subject to treatment as new sources, as long as emissions are not predicted to increase.

Only if they modernize in ways that increase emissions (for example, by increasing the plant's productive capacity), do they risk having to comply with strict new source controls. But from an economic dynamic perspective, one might want to encourage the modernization choice whenever equipment must be replaced or even repaired. This might suggest that Congress should apply new source standards to all renovations, rather than allow like-kind replacement of existing equipment through exemption from new source standards.

These modification regulations may have had an enormous impact upon air quality in practice. Recently, New York announced a suit against power plants in several upwind states. A little later, the Justice Department, several eastern states, and a few environmental groups announced plans to either join New York's suit or file their own suits against power plants in a number of other states. These power plants emit enormous quantities of nitrogen oxide from coal-fired power plants. Nitrogen oxides contribute in a major way to ozone (smog), particulate (which is positively correlated with tens of thousands of annual deaths), acid rain, and to climate change. All of these plaintiffs allege that the plants' owners violated the CAA requirements applicable to new and modified sources. The lawsuits claim that these owners modernized their plants to increase emissions without applying for the required permit for new sources. In other words, the governments and environmental groups claim that the plants evaded strict new source control requirements. The plant owners, however, argue that the projects involved like-kind replacement of old equipment that would not increase emissions. If this claim is correct, then the plant owners faced no requirement to meet new source standards.

This suit shows that significant polluters have renovated without new source controls. The courts will have to decide whether they did so because the projects were properly exempt under the regulations, or because the operators engaged in illegal evasion of the regulations.[11] But in either case, it is clear that, in practice, renovations have gone on for many years without upgrades in pollution control.

This presents an economic dynamic complication for those who blame the strict new source standards for the lack of turnover in equipment. First, there has been turnover in equipment. Second, if the projects chosen did not modernize in a way helpful to air quality one must evaluate several competing explanations. One explanation, the conventional one, would claim that the stricter standards caused plants not to modernize. Since plants may modernize on-site as long as emissions remain constant or decrease without new source controls, this explanation has some defects. The other explanation would state that the lack of requirements for all renovation projects to meet strict pollution control standards provides a loophole that allows plants to continue running without meeting new source standards. A third possible explanation is that the regulatory system has too little impact to provide any meaningful incentives affecting modernization decisions. One must seriously analyze the economic dynamics and corporate decision-making to figure this out.

The answers to these questions are important because EPA has been engaged in years of discussion about how to "reform" new source review. The answers to the questions posed here would contribute to the ongoing debate.

This debate has become especially prominent recently, because of anxiety about high energy prices in California. President Bush announced that he planned to encourage new plant construction to supplement existing power supply in response to the California problem. To facilitate this new construction, he promised to "streamline" government approval. In other words, he planned to relax new source review. This apparently reflected a conclusion that environmental regulation had prevented modernization.

In objecting to this relaxation, environmentalists claimed that new source review had not stopped the siting of power plants. Instead, industry had simply not sought to build new power plants because, until recently, we appeared to have too much power generating capacity. Furthermore, as it became clear in California that more power plants might be desirable, proposals had moved forward. This is obviously an important claim that the free market incentives have been more important then the regulatory incentives so many associate with new source review. Obviously, thoughtful, rather than reflexive, evaluation of these competing claims would aid evaluation of crucial policy questions.

Professor Bruce Ackerman of the Yale Law School and Professor Richard B. Stewart have recommended emissions trading as a cure to the alleged problem of new source standards discouraging modernization.[12] Yet Professor Stewart points out that one can enhance the political acceptability of emissions trading by grandfathering in the emissions of existing sources, meaning allowing existing sources to continue current high rates of pollution.[13] If an emissions trading program allows existing polluters to continue polluting at existing rates, it can only prevent increases by not allowing emissions from new sources, absent the purchase of credits. Emissions trading can either favor or disfavor strict regulation of new sources. Clearly, the choice of emissions trading by itself does not solve the problem of discouraging modernization. Rather, choices about how to design trading will control how trading addresses new sources, just as decisions about how to design traditional regulation influence the treatment of new sources.

Imagine an emissions trading program designed as follows. Congress passes a law stating that no electric power plant may operate without an allowance for each ton of nitrogen oxide it emits. It then gives allowances to each plant representing 50 percent of each plant's current emissions, effectively requiring reductions from each plant. An owner of a new facility would have to purchase allowances from existing companies under this scenario. And in a deregulated electric utility industry, existing utilities may want to discourage the competition by not selling their allowances to new entrants. Obviously, such a trading system would discourage new facilities and the accompanying advancement of technology.

Similarly, if the government auctions off a limited supply of allowances, as Professors Ackerman and Stewart have proposed,[14] existing facilities with large revenues may outbid fledgling competitors. This does not mean that emissions trading, with or without auctions, is a bad idea. Nor does it mean that one cannot design trading programs to favor new sources. It does mean, however, that in both the context of traditional regulation and in the context of trading, design choices, not the choice between traditional regulation and trading, determine how new sources are treated.

Economic dynamic analysis of new source review would benefit from more careful scrutiny of how regulatory considerations figure in private decision-making. Stating that a regulatory program poses a barrier to innovation because it imposes strict requirements is not enough. One must analyze how private decision-making proceeds under alternative scenarios. And one must carefully analyze how specific competing regulatory designs might affect private decisions.

Regulatory Design

The issue of regulatory design has received amazingly little attention. Part of this failure to think about this important issue comes from the simplistic preoccupation with the project of bashing "command and control" regulation and promoting emissions trading. Just as either traditional regulation or emissions trading can promote modernization projects or discourage them, depending on how they treat new sources, the design of either emissions trading or traditional regulation can influence economic dynamics more generally. This should not be surprising. Emissions trading begins as a traditional regulation limiting emissions and then contains an authorization to trade. While the trading may enhance regulation's cost savings, the design of the regulatory limits that motivate the trading will influence a trading program, just as design influences a program that does not authorize trading.

Environmental Metrics

Agencies face choices about how to write pollution limits. The metric employed influences the dynamics of regulation, traditional regulation and emissions trading regulation. Yet few analysts have thought about the policy implications of environmental metrics.[15]

Suppose that an agency decides to mandate a fixed percentage reduction from polluters' existing baseline emission levels. A fixed percentage reduction requirement implies that plants that have already achieved low pollution levels will have to reduce pollution even more. Plants that have high pollution levels face a laxer absolute limit on emissions. For example, assume that one plant now emits 100 tons of a pollutant a year and the other emits 200 tons. A 50 percent reduction requirement would force the 200 ton emitter down to a 100 ton per year level and the 100 ton emitter down to a much lower level—50 tons per year.

Use of percentage reduction requirements over time sends a message to companies. It suggests that innovating to reduce pollution can put one in a box where one must meet exceedingly tough demands. It makes it easier to keep high pollution levels to make substantial reductions easier.

A similar economic dynamic exists when agencies require a percentage reduction from uncontrolled levels. This occurs quite frequently, because one can set such a level without knowing a lot about particular baseline emissions in an industry. A regulatory agency that knows that a given end-of-pipe technology can reduce emissions by 90 percent, for example, can mandate that reduction without knowing much about the baseline emissions. Notice, however, that this choice does not reward companies that reduce their baseline emissions, for example, through pollution prevention. They have lower uncontrolled levels, but they must achieve a fixed percentage reduction from this lowered level. Their competitors need to make a fixed percentage reduction from a higher level. This means that the real level of emissions will be lower for the polluter that already made some improvement. The percentage reduction requirement is equally easy for all companies to achieve.

On the other hand, the government may require fixed levels of pollution. With fixed levels, polluters that have high pollution levels may have more difficulty in meeting these standards than those who have already achieved lower levels of pollution. Fixed levels generally provide for a more positive economic dynamic. They may reward those who have come closest to the required level and prove more difficult for those with higher baseline pollution levels.

These levels, however, may prove more difficult for an agency to set, if it has a goal of making sure that all companies can comply. Setting an absolute level may prove difficult for companies with very high emissions. Evaluating whether they can meet that level requires a knowledge of baseline emissions as well as the techniques available to reduce emissions.

The relative economic dynamics of absolute levels and percentage reduction requirements matter both to traditional regulation and emissions trading. Agencies have sometimes designed trading programs based on percentage reduction requirements. The pollution level a company calculates by applying a mandatory percentage reduction to baseline emissions becomes the allowance for all practical purposes. It may sell reductions made below this level and must purchase credits in order to exceed this number. This implies that companies with high current emission get a lot of allowances. This method sends a negative economic dynamic signal over time, just as it would in a traditional regulatory program.

In the acid rain program, Congress distributed allowances reflecting fixed levels of pollution. This was possible because EPA had studied the electric utility industry for many years and actually knew their emission baselines. Normally, EPA does not really know that much about industry emissions.

EPA has also made important policy choices regarding the metric used to express either a reduction requirement or a fixed level. I will focus on the fixed level case for simplicity's sake.

Many regulations, including both traditional regulation and state emissions trading programs, limit emission rates.[16] Emission rates limit the amount of pollution per unit of activity. For example, many regulations limiting air pollution coming from applications of paints, coatings, and solvents limit the pounds of emissions per gallon of substance used. EPA has traditionally regulated electric utilities through limits on the pounds of pollution per Million British Thermal Units (MMBTUs).[17] BTUs measure energy use.

The federal acid rain program and the federal regulations implementing the phaseout of ozone depleting substances, however, actually limit the mass of permitted pollution. The CFC (chlorofluorocarbon) regulations limited the tons of substances produced per year. The acid rain program limits tons of sulfur dioxide emitted per year.

This distinction between mass-based and rate-based limits matters a lot to the effectiveness of regulation and its economic dynamic. A rate-based regulation does not limit the mass of pollution that a pollution source may discharge. If a company's activity levels increases, so will its pollution. On the other hand, a mass-based regulation limits the actual quantity of pollution allowed. If a company wishes to increase its production, it must reduce its emissions rate so as to meet the mass-based requirement.

This means that mass-based limits provide a built-in economic dynamic that rate-based limits lack. A company wishing to produce more of a product to meet rising demand must find ways to obtain further pollution reductions in order to meet a mass-based cap. By contrast, a company subject to a rate-based emission level can increase production and therefore pollution levels and remain in compliance with its limits without further improvements or environmental innovations.

This difference matters in the emissions trading context as well. A company subject to a rate-based limit need not purchase credits when its activity levels increase in a trading program based on emission rates. Under a mass-based emissions trading program, such as the acid rain program, companies that wish to increase their activity levels must, at some point, either make additional reductions or purchase additional reduction credits. This means that more of an incentive exists for environmental improvement and, hence, environmental innovation, under a mass-based trading program than under a rate-based trading program.

One might ask why the regulatory system relies so heavily upon rate-based limits, if mass-based limits offer superior incentives for innovation and environmental performance. Mass-based limits make economic growth more difficult for regulated companies. Increased production under a mass-based limit triggers a need for additional pollution prevention. This suggests that the economic dynamics of the regulatory process, and similar dynamics in Congress, might tend to favor rate-based limits.

Greater reliance on mass-based limits, however, meets a number of important goals relevant to economic dynamics. Rate-based limits increase the need for repeated regulatory decisions, since they allow emission increases, even in regulated sectors, as the economy grows. This increases the risk of the regulatory system not keeping up with the pace of environmental destruction. It also increases the complexity of environmental law by creating a need for multiple regulations to achieve a single goal. Rate-based limits require the regulatory decision-making structure, rather than the quicker, more efficient private market decision-making structure, to respond to the environmental problems associated with increased production.

Mass-based limits also provide a tool to help implement the concept of sustainable development as developed by Herman Daly. Recall that Daly favors economic development without increased "throughput" of resources. But translating even Daly's relatively precise ideas relating to sustainable development into a set of steps that a regulatory system can hope to execute poses an enormous challenge. The idea of avoiding increased throughput in order to foster economic development raises a fundamental question—how should this idea influence the law of environmental protection? Mass-based limits provide a piece of the answer. Mass-based regulations limit throughput. Their implementation provides a concrete step toward sustainable development, whether executed with or without allowance trading.

Adoption of mass-based limits will require government to change how it thinks about environmental protection. Government must have some faith in private sector capacity to carry out environmental innovation in order to adopt mass-based limits. And it must be willing to assign the private sector, rather than government, the responsibility to reconcile environmental goals with economic development.

Another key variable affecting innovation is stringency.[18] Regulations have sometimes produced substantial innovations.[19] These regulations include bans on ozone depleters, state rules requiring zero emission vehicles, the phaseout of lead from gasoline, and some of the stricter standards regulating occupational health and safety. All of them have one thing in common—they impose stringent standards on pollution sources.[20]

At first glance this might seem paradoxical. After all, stringent regulation narrows the choice of technological possibilities. To take an extreme example, suppose that a regulation requires no reductions at all from an industry. This leaves the regulated company free to adopt any technology it wishes regardless of pollution characteristics. It may do nothing, employ end-of-the-pipe controls, engage in pollution prevention, or completely change its production process. The economic dynamics, however, will tend to favor the do nothing option. So, while lax regulation makes a wider range of technological options legally available, it makes adoption of environmental innovations costing any money at all unlikely.

Mediocre regulation, the kind that dominates much of environmental law, requires some improvement, usually based on end-of-the-pipe controls that an agency believes is reasonably cost effective. When an agency goes beyond mediocre regulation and demands enough reduction to make business as usual either impossible or very expensive, private parties have an incentive to innovate to escape the high cost. Stringent regulation has brought about the elimination of lead from gasoline, the sale of substitutes for CFCs, the removal of some regulated toxic chemicals from some occupational settings, and other rather significant changes. By limiting opportunities to use standard technologies, stringent regulation encourages innovation. Analysts regularly confuse opportunities for a wide range of technological response (such as provided by the absence of an environmental regulation) with a substantial incentive to provide environmental innovation.

Substantial barriers exist to stringent regulation. A major problem at the federal level has been EPA's tendency to assume the burden of proving that it knows how a regulated industry will meet an environmental target, rather than leaving the primary business of figuring that out to industry.[21] For example, EPA promulgated a new source performance standard for lime kilns in the 1980s. During the rulemaking it interpreted the relevant statutory provisions to require it to have a reasonable basis to conclude that lime kilns could meet the limits using the full variety of feedstocks currently employed in the industry. A court invalidated its regulation, because the agency failed to show how its limited sample of emission control tests demonstrated that its technology would work adequately for all feedstocks and other operational variables.[22] The court's decision, however, rested upon the agency's position assuming the burden of showing that all plants using all feedstocks could meet the limit. The agency did not have to take that position. Indeed, its position does not fit the economic dynamic rationale for new source standards. The Act's new source provisions are designed to assure that turnover produces really good emissions performance, not just environmentally mediocre choices. The agency could have simply pointed out that companies building new plants may choose which feedstocks to rely upon and design their plants accordingly. Since the companies can control this and other operational variables, the agency could reasonably conclude that a showing that the emissions limitations can be met under any set of operational circumstances suffices to show that the regulation is achievable.

Perhaps the agency failed to take this position because new source performance standards apply to some renovations of existing facilities, not just brand new plants. This raises the prospect of new source performance standards forcing existing plants with aging equipment to change their feedstocks and operational methods. An economic dynamic perspective, however, suggests that the law should not view the possibility of putting pressure for change on owners of existing facilities negatively. First, facility owners may have some capacity to make changes. Second, if they do not, perhaps they should lose out to new facilities.

EPA has taken positions comparable to its National Lime position even under statutory provisions specifically designed to encourage more stringent standard setting. For example, the provisions regulating air toxics under the 1990 Amendments include a requirement that all existing major sources of regulated air toxics meet the average performance level of the best performing 12 percent of existing sources. Congress wrote this limit precisely to overcome the lack of stringency in the program that produced the lime kiln regulations. Yet EPA has assumed the same kind of burden in adopting these standards, instead of simply noting that the statute allows the agency to assume that most plants can do what the best plants can do.[23]

This whole problem of stringency also flags a general issue, EPA's reluctance to make any decision that substantially limits operational flexibility. Some industries tend to like end-of-pipe controls because they require little imagination and do not disturb their basic production process. EPA has responded by largely acquiescing to industry desires. Perhaps the most serious example of this comes from the electric utility industry. Electric utilities use boilers capable of running on a variety of fuels, including coal and natural gas. Historically, EPA has written different standards for each fuel type. So instead of insisting on environmental performance reflecting what can be achieved with the most environmentally friendly fuel choice, it has simply assumed that fuel choice should remain free of federal environmental influence.

EPA's latest revision of the new source performance standards for electric utilities signals that this may be changing. In 1998, EPA adopted a fuel neutral standard for electric utility emissions.[24] But this standard is too lax to require fuel switching.[25]

To be fair to EPA, a decision forcing companies to forego coal in favor of natural gas would threaten coal miners' jobs and might trigger a severe Congressional reaction. Furthermore, the courts sometimes place great burdens upon agencies trying to force technologies, even if the agencies take legal positions that facilitate technology forcing rulemaking.[26] These political and legal pressures, however, simply highlight the relationship between the economic dynamic analysis of decision-making processes and the economics of regulatory design. The strong influence of existing industries upon decision-making tends to make it hard to adopt economically dynamic regulatory design, so reform of decision-making structures may be a political prerequisite to design reform.

Nevertheless, stringent regulation might be an option. One might contrast EPA's performance in recent years with that of the California Air Resources Board (CARB). CARB has introduced some technology forcing changes, most notably, perhaps, its low emission vehicle program.[27] As initially conceived, this program required reductions in the short term achievable through fairly minor innovations, with a requirement for introduction of much cleaner vehicles, including a small number of zero emission vehicles, several years in the future.[28] While CARB has relaxed its program somewhat under pressure from the auto industry, the LEV program has nevertheless proved a very important catalyst to development of cleaner technologies in the car industry. Car companies have moved forward in developing fuel cells and advanced battery technologies in response to the LEV program, which several eastern states have moved to implement as well. A zero emission vehicle would indeed be a transformative technology, delivering enormous environmental benefits.

CARB was able to do this for a variety of reasons. CARB has funded a laboratory specializing in advanced technologies. This knowledge helped give it the confidence to impose stringent requirements with some confidence in their feasability, notwithstanding vehement claims to the contrary by much of the auto industry.

A more economic dynamic alternative to making agencies into superior technologists might be for agencies to stop seeing themselves as engineers and start seeing themselves as the source of demand for environmental technologies. They should set standards based on environmental needs, rather than cost or feasibility. This would require, however, some faith in the ability of industry to innovate to reduce costs and deliver environmental improvements, just as it introduces material innovations to meet consumer demand. This faith involves a wholesale shift in thinking. We must think of costs not as fixed obstacles, but as subject to change under pressure to meet demands. This kind of thinking could improve the design of both traditional regulation and emissions trading programs.

In theory, an agency can do a lot to enhance the economic dynamics of regulation. It can write regulations addressing broad categories of sources that force all to improve and encourage many to innovate to at least match the achievements of those who have made fundamental design decisions that favor the environment. But innovation and change imply disruption. A well financed army of lobbyists is available to fight changes threatening the status quo.

An economic dynamic analysis highlights how one could use regulatory design to improve the economic dynamics of environmental law. An understanding of the economic dynamics of the legal process governing environmental law raises some questions about whether government will in practice make the economically dynamic changes needed reasonably often. The privatization piece of the analysis provides an alternative for those skeptical of government's willingness or ability to write economically dynamic regulation in spite of the pressures upon them.

In general, then, economic dynamic analysis raises the issue of how to improve the economic dynamics of environmental law. This leads to a sharply different set of questions than those currently dominating the field. These question include the question of privatization, of making administrative procedure more fair and effective, and of how to improve regulatory design. Focusing the policy debate upon these questions would constitute a significant and worthwhile change.

IV
Toward Economic Dynamic Legal Theory

This part explains why economic dynamic thinking should reshape law and economics more generally. It accomplishes this both by drawing some broad lessons from the prior analysis of the economic dynamics of environmental law and by applying economic dynamic concepts to another area, that of regulated industries. This part ends with a short recapitulation of the book's principal conclusions.

12

Static Efficiency Reconsidered

This book explains that efficiency-based thinking has dominated the debate about regulatory reform in the environmental area. It claims that economic dynamic analysis offers an alternative to efficiency-based thinking and raises important questions about the efficiency-based regulatory reform agenda. This chapter assesses economic dynamic's more general implications for efficiency-based analysis. This assessment highlights economic dynamic's general implications for legal theory and shows that economic dynamic analysis will prove important even outside the environmental area.

While this book has focused upon the economic dynamics of environmental law, a description of economic dynamics analysis in more general terms will make its wider potential more apparent. Economic dynamics analysis, building on the idea of path dependence, begins with noticing what types of economic changes will appear worthwhile over time to private actors of special importance to the legal regime under consideration. The analysis continues by figuring out what desirable changes will not appear worthy of pursuit by these private actors. Some analysis of the ways emerging technologies may affect the values at stake also should play a role. Analysis of the economics to determine who will profit from future innovations will tell us who has the financial capability to invest a lot of money in lawyers and lobbyists to shape governmental regulation and what changes they will likely seek. Furthermore, understanding the relevant private actors and their orientation will aid evaluation of the consequences of legal rules.

Beyond Environmental Law: The Example of Regulated Industries

Economic dynamic analysis of this kind will prove important in areas widely thought to be the most fruitful areas for application of efficiency-based analysis. Consider regulated industries. For many years, the government regulated common carriers and public utility companies to assure non-discriminatory pricing and equal access to essential services.[1] The standard law and economics story involves deregulation improving the efficiency of these formally regulated industries.[2] It would require another book to evaluate whether deregulation has improved efficiency in the technical sense and to thoroughly compare efficiency improvements to economic dynamic changes. Most reviews so far have concluded that deregulation has generally generated large average cost savings, thereby improving average consumer welfare.[3] But economic dynamic issues matter a lot to the future of the law of regulated industries, which suggests that the ideas outlined in this book will bear fruit in other areas.

Sophisticated analysis of regulated industry law adopts a nuanced characterization of the transformation of the law of regulated industries and the issues that the new law must grapple with.[4] If "deregulation" means the elimination of all public regulation of utilities and common carriers, then deregulation has not occurred. But a radical change in regulation has occurred, which features greater reliance upon competition, an element that this book has emphasized as an important part of free market dynamics, and less reliance upon government rate-setting.[5]

Most of the interesting questions that this change raises will require economic dynamic analysis. And the results of that analysis will drive future policy.

For example, an equitable issue, whether a regime featuring competition can deliver reasonably priced services for low income and rural consumers, libraries, and educational institutions will play a key role in the future of telecommunications. While the Telecommunications Act of 1996 envisions less direct government regulation of the rates and services telecommunications companies provide than in the past, it contains a provision strongly supporting a policy of universal service.[6] Indeed, the Act requires discount services for "educational providers and libraries."[7]

A key issue emerges—can universal service survive a regime based upon competition?[8] Notice that the universal service ideal in this deregulatory statute conflicts with efficiency, at least to some degree. For example, it may cost more to provide services in a rural area than in an urban area. Efficiency would seem to favor having good service in urban areas and poor or no service in rural areas.[9] But the Telecommunications Act rejects this solution.

The question of whether universal service will survive the new competitive environment involves economic dynamics. In particular, it requires an analysis of how the incentives in the statute might influence telecommunications providers. This analysis might focus upon what directions in the law these companies will tend to favor, which will flow from an analysis of how they can maximize their profits. Any telecommunications provider that charges businesses enough money to provide a subsidy for rural residential services (a cross-subsidy) risks having a competitor steal the market for services to businesses.[10] A competitor has an incentive to provide the services for business at a lower price by not charging enough to cross-subsidize services in rural areas. The Telecommunications Act contemplates charging a fee to provide the cross-subsidies necessary to meet the universal access goal.[11] Professors Joseph D. Kearney of Marquette University Law School and Thomas W. Merrill of Northwestern University School of Law point out that the Act gives long-distance service providers and those customers paying for the cross-subsidies incentives to keep the fees funding them low.[12] A key question then involves an economic dynamic problem of who will dominate rulemaking proceedings—the beneficiaries of subsidies or the payers. The economic dynamic theory outlined previously gives some ways of thinking about that problem and economic dynamic reforms (like the reforms of the administrative process discussed in this book) would affect the outcome of a key issue.

Similar equitable problems exist in other regulated industries. Currently, deregulation of utilities has principally benefitted large business customers.[13] A key issue remains whether deregulation will harm or help small business and individual consumers.

In California, electricity prices rose greatly shortly after deregulation. A key issue in the debate about how to respond to this involves economic dynamics. California forbade electric utilities from buying power with long-term contracts, thereby forcing them onto the spot market. This might create an incentive for power suppliers to withhold power in hopes of driving up the prices on the spot market, for this strategy can increase their profits. If this economic dynamic explains the problem, then some adjustment in regulation must remove this incentive.

President Bush, however, has defined the problem in classic efficiency terms, claiming that a supply shortage lies at the root of the problem. From that perspective, the price increases are efficient, reflecting a proper matching of supply and demand.

This might suggest that no regulatory action limiting prices is needed, since the rising prices should provide incentives for more conservation. Such conservation should reduce demand. This is an application of the "induced innovation" hypothesis from economics, that the rising price of a production factor will spur innovation conserving the costly factor.[14] Focusing upon change over time in response to changing conditions, the principal focus of economic dynamics analysis, leads to identifying the precise link between this standard economic description and the important public policy question—how to address the rising prices. Reduced demand should bring down prices, so no action is necessary on this analysis.

In fact, however, as prices rose, energy producers began proposing to build new plants. So, the high prices might induce increased production, not reduced demand. This increased supply, however, would also tend to lower prices.

From the standpoint of efficiency, it does not matter whether demand decreases or supply increases. Either way, prices fall and equilibrium is restored. From the standpoint of environmental protection, it should matter a great deal. Increased supply means more pollution, decreased demand means less.

One might want to analyze how the incentives influence actual consumer and producer behavior to figure this out. Again, the economic dynamic approach, which analyzes the relationship between an incentive and the bounded rationality of potential actors, will help.

Technological innovation will also play a role in shaping the future path of economic regulation. This implies that the economic dynamics of innovation matter in this context.

Many (but not all) regulated industries have been regarded as natural monopolies, because of the technology employed in the industry. For example, electric power requires a network of wires to people's homes. Since stringing multiple sets of wires to people's homes would prove extremely inefficient, provision of electricity has been understood as a natural monopoly. For similar reasons, telephone service has also been regarded as a natural monopoly.

Changes in switching technology and computers made it easier for telephone companies to switch calls to the systems of rivals.[15] This change facilitated the move from monopoly to competition in the telephone system.

Government continues to regulate natural monopolies, and most sophisticated observers believe that such regulation is necessary in order to avoid inefficient monopoly. Thus, even under "deregulation," government requires local telephone exchange companies to file tariffs and has, so far, kept some tariffs in place in the natural gas and electric industries.[16] Even though some of these tariffs will probably disappear as "deregulation" proceeds, even under the newer laws in this area, regulators have a crucial role to play because of the economic dynamics of bottlenecks. Absent some government regulation, owners of essential pieces of services, such as transmission lines and telephone wires, would probably prevent delivery of services that might come over this infrastructure, or charge exorbitant monopoly rents to competitors. To assure competition, regulators have established an elaborate set of rules to prevent exploitation of these bottlenecks to stifle competition.[17]

In this area, study of the economic dynamics of competition, which considers how economic actors seeking to maximize their profits within the constraints imposed by their institutional customs, becomes a prerequisite to any effort to pursue efficiency. Thorough deregulation in the face of an effective bottleneck will not produce efficiency, it will produce monopoly. Since competitors have an incentive to work around bottlenecks, however, then the path of technological change matters to this analysis. For if competitors can and will introduce technologies circumventing bottlenecks, they will matter less.

Even those who disagree with the conclusion that government regulation is better than monopoly power do so for economic dynamic reasons. University of Chicago economist Milton Friedman's argument against regulation of natural monopolies assumes that regulated monopolies would so thoroughly capture the regulatory process as to exercise a de jure monopoly.[18] I have suggested that if such capture exists (and even if it doesn't), an economic dynamic helps explain regulated industry's influence. Furthermore, Professor Friedman makes another economic dynamic argument—

that this capture will lead the government to stifle future competition that might threaten the monopoly. Assessment of that argument requires some understanding of the bounds of bureaucratic rationality. Do bureaucrats so thoroughly identify with the industries that they regulate that they will stifle competition to protect monopolists? In other words, one cannot assess Friedman's argument without social science, in particular, studies of the institutional environment and constraints that influence the relevant government institutions.

While most mainstream economists have rejected the Chicago School position on this, another theory about market dynamics has proved very influential with congressional staff and regulators of utilities and common carriers.[19] William Baumol, a Princeton University economist also known for thoughtful writing about environmental economics, developed a theory of contestable markets, suggesting that lowering barriers to entry could produce efficiency gains.[20] This theory supported combining competition with regulation assuring access to bottleneck facilities at neutral prices.

Bottlenecks, however, may disappear as technology changes. For example, wireless transmission of telephone calls, if reliable and cheap enough, might raise questions about whether possession of telephone wires really involves a natural monopoly.[21] Figuring that out will require an analysis of competitive dynamics of the industry.

In short, economic dynamics sometimes aids design of legal institutions pursuing efficiency goals. It also helps analyze design issues in pursuing competing goals (such as equal access).

Competing Efficiencies

Some of this book's analysis can be reframed in efficiency-based terms. Doing this exposes another problem with efficiency-based analysis—the problem of conflicting efficiencies.

When we speak of efficiency in analytically meaningful terms we must specify what precisely we hope to maximize. For example, suppose that a person asks whether it is "efficient" to bike to work. No one can provide a clear answer to this question until the questioner specifies what variable she wishes to maximize. Does she mean to ask whether bicycling provides the least costly means of getting to work? Does she mean to ask whether bicycling provides the fastest way to work? Does she mean to ask whether it provides the most pleasant way to get to work? She must determine whether she wishes to maximize time, money, or pleasure before one can answer the question of whether bicycling is efficient.

This problem becomes even more acute where we have different actors involved in a transaction. For example, one might ask whether a business voice mail system is efficient. If one means to ask whether voice mail efficiently uses the service provider's financial resources, the answer may be yes, if the system substantially reduces labor costs, for example. But if one asks whether the system efficiently uses customer time, the answer may be no, if the system is complicated and cumbersome to use.

Regulatory reform focuses on efficient use of private sector compliance resources. The reforms recommended usually conflict with efficient use of government time or money devoted to enhancing environmental protection.

CBA provides the most egregious example of this conflict. It involves an enormous analytical effort that has frequently paralyzed government decision-making in the past. Emissions trading also involves some conflict between governmental and private sector efficiency. Monitoring allowance trading simply involves more government resources than monitoring compliance with an identically designed traditional regulation. Without allowance trading, government can check any given polluter's compliance by simply verifying that the polluter has complied with the applicable limits. With trading this does not suffice. If the source seems out of compliance, the regulator must check to see if purchased credits cover the exceedance. The regulator must then verify that the credit generating activity also performed as advertised when the credit was sold. In other words, verifying compliance at one facility requires monitoring emissions at several facilities under trading.[22] This makes spot checks, important deterrents, potentially cumbersome.

This problem of conflicting efficiencies implies that the justifications offered for efficiency-based reform are inadequate. Efficiency-based analysis must explain why one should maximize the analyst's preferred variable instead of another. Efficiency-based analysts have not explained why they prefer to minimize private sector compliance expenditures, rather than outlays of government resources devoted to crafting and enforcing environmental regulation. Indeed, they rarely even acknowledge the tension. They have, by and large, ducked rather than met the challenge of competing efficiencies. Economic dynamic theory's concern for the pace of decision-making and innovation highlights the existence of competing efficiencies.

An efficiency devotee might respond to this by pointing out that one could account for the agency time and expense along with private sector efficiency in evaluating regulatory reform.[23] In some sense, this is correct. Analysts must think about both private sector costs and public sector time and money in evaluating regulatory reform. But any serious thinking about how one should evaluate the public sector factors requires reliance upon analysis of factors that the economic dynamic analysis highlights.

For example, since CBA would slow agencies down tremendously, one should evaluate CBA as including a silent decision to leave many sources of pollution unregulated for a long period of time—a public opportunity cost (adapting the language of economists). One needs an analysis like that provided in the chapter on the pace of environment change to try and figure out whether this will matter a lot in the long run, and indeed, to even see the issue.

Similarly, any serious effort to evaluate the efficiency of emissions trading requires use of economic dynamics. If one concludes, as one should for the acid rain trading program, that the extra enforcement costs associated with trading are not terribly great, little problem exists in combining public and private concerns. But what if one determines that a particular trading program will make it very difficult for government to keep track of emissions? One needs to examine the economic dynamics (how polluters might behave if their emissions are inadequately monitored) to think about whether this will increase the emissions the trading program regulates. In the end, since the benefits and costs cannot be precisely measured and can only be properly identified with some underlying understanding of dynamics, stating that one must consider all of the costs really says very little.

Efficiency Considerations in Economic Dynamic Reform

My statement that efficiency has some value may lead one to ask whether economic dynamic reforms would undermine efforts to make environmental regulation's use of private sector compliance resources more efficient. Economic dynamic reforms could take cost and efficiency into account. I have explained, for example, that even the environmental competition statute that I outlined, a system designed to propel environmental innovation forward, could include review mechanisms to evaluate the costs of innovation. One could also build cost constraints into the statute a priori, albeit at some cost to its ability to spur innovation.

My suggestion that efficiency has received perhaps too much attention does raise an issue as to whether we should give yet more weight to cost considerations than we do now. But one can answer that question in one of several ways and still favor improved economic dynamics. The answer may influence design of reform, but it does not preclude economic dynamic reform altogether.

Some of the economic dynamic reforms suggested would directly enhance efficiency. For example, the existing legal system regularly overestimates cost. This overestimation may lead to inefficient regulations, i.e., regulations that are too lax. Optimal regulation has costs equaling benefits, but a number of regulations allow pollution that could be controlled more stringently without causing an imbalance. If equalizing the lobbying resources of environmental groups, entrepreneurs, and industry would ameliorate this tendency to overestimate cost, this should make environmental regulation better approximate its goals, whatever they might be. If the goal were efficiency, more balanced representation should improve the accuracy of regulatory estimation of costs and benefits at the heart of a cost-benefit system.

Economic dynamic analysis raises crucial questions about the value of efficiency-based analysis and reform. One can pursue economic dynamic reform with little regard for private sector costs, or in ways that tend to enhance, or at least not harm, efforts to reduce private sector compliance expenditures. Consideration of economic dynamics aids evaluation of questions about whether cost considerations need more or less attention. It points out the need to consider how proposals that adjust regulatory processes to reduce private sector compliance costs will affect innovation over time. This assessment must include serious consideration of the future need for environmental protection.

13

Conclusion

Consideration of economic dynamics should be central to analysis and reform of environmental law. The theory of economic dynamics reframes analysis, raises significant questions that have received too little attention, and introduces new possibilities for reforming environmental law.

Economic dynamic theory calls our attention to the temporal dimension of environmental protection. It asks how well environmental law can cope with growing population, innovation, and economic growth over time. This inquiry leads to new modes of analysis. Serious consideration of the role of innovation in changing the shape of our society over time becomes important. It becomes relevant to compare the economic incentives for innovations increasing material well-being to those for environmental innovation. A comparative analysis of private and public decision-making structures governing response to economic incentives becomes significant.

This analysis then leads to new questions about environmental law. Scholars and policy-makers need to think about whether environmental law can and should be privatized. This question requires serious thought about what privatization might mean and how it might be accomplished. It leads to asking about how the process of environmental decision-making could be made more efficient and fair. And economic dynamics implies that analysis of legal process requires one to view decision-making processes as reflecting the work of outside forces, not as a self-contained system. Finally, economic dynamic theory requires more precise analysis of regulatory design issues. Proper analysis of regulatory design can support efforts to achieve sustainable development. This mode of analysis goes well beyond simplistic generic dichotomies damning traditional regulation as "command and control" regulation and lauding "economic incentive" measures.

It involves a more fruitful and more precise look at what can be improved and how. More generally, law and economics will benefit from greater attention to economic dynamic analysis.

We live in a dynamic, changing world, a world ill-suited to static analytical frameworks. Our way of thinking about environmental law and policy must change to meet the demands of the world we live in. Indeed, our way of thinking about most areas of law needs to change to reflect and respond to important economic dynamics.

Notes

Chapter 1

1. See, e.g., Richard A. Posner, *The Economics of Justice* (Cambridge, Mass.: Harvard University Press, 1983), 104, claiming that the public interest theory is not a theory, but an erroneous description; Jerry Mashaw, *Greed, Chaos, and Governance: Using Public Choice to Improve Public Law* (New Haven: Yale University Press, 1997), 23; "All things public have become suspect. For some, the only public purpose worthy of respect seems to be the elimination of the public sector."

2. For a very brief discussion of the influence of law and economics by one of its principle figures see Richard A. Posner, *Overcoming Law* (Cambridge, Mass.: Harvard University Press, 1995), 96–97. See also Roger C. Park, "Lawyers, Scholars and the 'Middle Ground,'" *Michigan Law Review* 91 (1993): 2075–2112, 2084–2085.

3. See Howard Latin, "Ideal Versus Real Regulatory Efficiency: Implementation of Uniform Standards and 'Fine-Tuning' Regulatory Reforms," *Stanford Law Review* 37 (1985): 1267–1332.

4. See generally Susan Rose-Ackerman, *Rethinking the Progressive Agenda: The Reform of the American Regulatory State* (New York: Free Press, 1992).

5. See generally E. S. Savas, *Privatizing the Public Sector: How to Shrink Government* (New York: Chatham House, 1982).

6. See David M. Driesen, "Getting Our Priorities Straight: One Strand of the Regulatory Reform Debate," *Environmental Law Reporter* 31 (2001): 10003–10020.

7. See A. Denny Ellerman et al., *Markets for Clean Air: The U.S. Acid Rain Program* (New York: Cambridge University Press, 2000), 312–313, noting common market imperfections.

8. Kenneth J. Arrow et al., *Benefit-Cost Analysis in Environmental, Health, and Safety Regulation: A Statement of Principles* (Washington, D.C.: American Enterprise Institute for Public Policy Research, 1996), 3.

9. See Posner, *The Economics of Justice*.

10. See Michael S. Common, *Sustainability and Policy: Limits to Economics* (New York: Cambridge University Press, 1995), 139; National Science and Technology Council, *Technology for Sustainable Development* (Washington, D.C.: National Science and Technology Council, 1994), 2. Technology has been responsible for as much as two-thirds of the increase in the nation's productivity since the Depression. Lewis M. Branscomb and Richard Florida, "Challenges to Technology Policy in a Changing World Economy," in *Investing in Innovation: Creating a Research and Innovation Policy that Works,* ed. Lewis M. Branscomb and James H. Keller (Cambridge, Mass.: MIT Press, 1998), 41. Advances in "technical know-how" probably account for one quarter to one half of total U.S. economic growth since World War II.

11. See generally Robin Paul Malloy, *Law and Market Economy: Reinterpreting the Values of Law and Economics* (Cambridge: Cambridge University Press, 2000), 78–99, 137.

12. See John Kenneth Galbraith, *American Capitalism: The Concept of Countervailing Power* (Boston: Houghton Mifflin, 1952); Joseph Alois Schumpeter, *Capitalism, Socialism, and Democracy* (New York: Harper, 1947); J. S. Mill, *Principles of Political Economy,* Book IV, Ch. VII (London: John W. Parker, 1852), 352; John B. Clark, *Essentials of Economic Theory* (New York: Macmillan, 1907), 374.

13. See F. M. Sherer, "Schumpeter and Plausible Capitalism," reprinted in *The Economics of Technical Change,* ed. Edwin Mansfield and Elizabeth Mansfield (Brookfield, Vt.: Elgar, 1993), 183. Incentives for innovation are almost surely inadequate in a world of atomistic competitive markets, but doubt has been cast on Schumpeter's view that large firms innovate more than small ones.

14. See Malloy, *Law and Market Economy,* 32–33. New information, chance, and surprise destabilize the observed equilibrium through an evolutionary and dynamic process.

15. See *Dynamics, Economic Growth, and International Trade,* ed. Bjarne S. Jensen and Kar-yiu Wong (Ann Arbor, Mich.: University of Michigan Press, 1997), 31, noting the positive correlation between technological improvement and growth.

16. See Malloy, *Law and Market Economy,* 85.

17. See Scherer, "Schumpeter and Plausible Capitalism."

18. See Douglass C. North, *Institutions, Institutional Change, and Economic Performance* (New York: Cambridge University Press, 1990), 81. Adaptive efficiency and allocative efficiency may not always be consistent.

19. See Oliver Williamson, ed., *Organization Theory: From Chester Barnard to the Present and Beyond* (New York: Oxford University Press, 1995), 106–107, 178–179, 185, 188–191.

20. Environmental Law Institute, *Cleaner Power: The Benefits and Costs of Moving from Coal Generation to Modern Power Technologies* (Washington, D.C.: Environmental Law Institute, 2001), 4–5.

21. See North, *Institutions,* 80.

22. Ibid.

23. Ibid., 81.

24. Ibid., 99.

25. See generally Mashaw, *Greed, Chaos, and Governance.*

26. See Herman E. Daly, *Beyond Growth: The Economics of Sustainable Development* (Boston: Beacon Press, 1996), 65, 194–195; Nicholas Georgescu-Roegen, *The Entropy Law and the Economic Process* (Cambridge, Mass.: Harvard University Press, 1971).

27. See Edward L. Rubin, "The New Legal Process, the Synthesis of Discourse, and the Microanalysis of Institutions," *Harvard Law Review* 109 (1996): 1393–1438; Oliver E. Williamson, "Why Law, Economics and Organization?," *UC Berkeley Public Law and Legal Theory Working Paper No. 37* (2001); David M. Driesen, "The Societal Cost of Environmental Protection: Beyond Administrative Cost-Benefit Analysis," *Ecology Law Quarterly* 24 (1997): 545–617, 563–577 (discussing macro-economic goals and sustainable development); Timothy Malloy, "Regulating by Incentives: Myths, Models and Micro-Markets," *University of Texas Law Review* 80 (2002): 531–605 (applying institutional economic analysis to the problem of economic incentives in environmental law); Douglas A. Kysar, "Sustainability, Distribution, and the Macroeconomic Analysis of Law," *Boston College Law Review* 43 (2001): 1–71, proposing that sustainability considerations should change law and economics.

Chapter 2

Much of chapter 2's analysis is based on a longer article, David M. Driesen, "The Societal Cost of Environmental Regulation: Beyond Administrative Cost-Benefit Analysis," *Ecology Law Quarterly* 24 (1997): 545–617.

1. See, e.g., Clean Air Act, U.S. Code, vol. 42, sec. 7412(d)(2) (1994)(supp. V); *Michigan v. Thomas,* 805 F.2d 176, 181 (6th Cir. 1986); *Union Electric v. EPA,* 427 U.S. 246, 266 (1976). States may consider costs in choosing strategies for meeting national ambient air quality standards.

2. See, e.g., *American Textile Manufacturers v. Donovan,* 452 U.S. 490, 540 (1981).

3. Federal Insecticide, Fungicide, and Rodenticide Act, U.S. Code, vol. 7, secs. 136–136(y) (2000).

4. Toxic Substances Control Act, U.S. Code, vol. 15, secs. 2601–2692 (2000).

5. John S. Applegate, "The Perils of Unreasonable Risk: Information Regulatory Policy and Toxic Substances Control," *Columbia Law Review* 91 (1991): 261–333 (FIFRA and TSCA use language that authorizes action to prevent "unreasonable" adverse effects and have legislative history calling for balancing of costs and benefits); Alan Rosenthal, George M. Gray, and John D. Graham, "Legislating Acceptable Cancer Risk from Exposure to Toxic Chemicals," *Ecology Law Quarterly* 19 (1990): 269–362, referring to FIFRA and TSCA as risk-balancing statutes; *Corrosion Proof Fittings v. EPA,* 947 F.2d 1201, 1217 (5th Cir. 1991),

describing statutory requirements of TSCA. The 104th Congress amended the legal regime governing pesticides to apply standards not based upon cost-benefit analysis to pesticide residues in food and pesticides preventing disease. See The Food Quality Protection Act of 1996, U.S. Statutes at Large 110 (1996): 1508, 1516–1517 (requiring health based standards for pesticide residues in food and risk/benefit standards for pesticides protecting public health). The experience prior to the 1996 amendments remains relevant to understanding CBA.

6. See President, Executive Order, "Executive Order 12,291," *Federal Register* 46, no. 33 (19 February 1981): 13193, requiring that benefits outweigh costs "to the extent permitted by law"; Eric D. Olson, "The Quiet Shift of Power: Office of Management and Budget Supervision of Environmental Protection Agency Rulemaking under Executive Order 12,291," *Virginia Journal of Natural Resources Law* 4 (1984): 1–80b, 25–27, citing cases; Heimann et al., "Project: The Impact of Cost-Benefit Analysis on Federal Administrative Law," *Administrative Law Review* 42 (1990): 545–654, 602; Alan B. Morrison, "OMB Interference with Agency Rulemaking: The Wrong Way to Write a Regulation," *Harvard Law Review* 99 (1986): 1059–1074, 1062. See also *Environmental Defense Fund v. Thomas*, 627 F. Supp. 566 (D.D.C. 1986), executive orders cannot justify ignoring statutory deadlines; Thomas O. McGarity, "Regulatory Analysis and Regulatory Reform," 65 *Texas Law Review* 1243–1333, 1319 (1987). "An agency cannot rely upon the Executive Order . . . to take unauthorized action or to refrain from taking required action."

7. Olson, "The Quiet Shift of Power," 49–55; Thomas O. McGarity, "Some Thoughts on Deossifying the Rulemaking Process," *Duke Law Journal* 41 (1992): 1385–1462, 1431–1433); Thomas O. McGarity and Sidney A. Shapiro, *Workers at Risk: The Failed Promise of the Occupational Safety and Health Administration* (Westport, Conn: Praeger, 1993), 229–241, summarizing OMB intervention detailed throughout the book.

8. See William J. Baumol and Wallace E. Oates, *The Theory of Environmental Policy* (Englewood Cliffs, N.J.: Prentice-Hall, 1975), 23.

9. See Driesen, "The Societal Cost of Environmental Regulation," 564.

10. See William F. Baxter, *People or Penguins: The Case for Optimal Pollution* (New York: Columbia University Press, 1974).

11. See E. J. Mishan, *Cost-Benefit Analysis* (Boston: G. Allen and Unwin, 1982), 162, referring to Kaldor-Hicks efficiency as a potential Pareto improvement. See also Lawrence A. Tribe, "Policy Science: Analysis or Ideology," *Philosophy and Public Affairs* 71 (1972): 66–110.

12. Justice Breyer has become perhaps the most prominent proponent of this view. See Stephen G. Breyer, *Breaking the Vicious Circle: Toward Effective Risk Regulation* (Cambridge, Mass.: Harvard University Press, 1993).

13. See, e.g., Cass R. Sunstein, "Legislative Forward: Congress, Constitutional Moments, and the Cost-Benefit State," *Stanford Law Review* 48 (1996): 247–309, 257–260.

14. See Lisa Heinzerling, "Regulatory Costs of Mythic Proportions," *Yale Law Journal* 107 (1998): 1981–2070.

15. See, e.g., Matthew D. Adler and Eric A. Posner, "Rethinking Cost-Benefit Analysis," *Yale Law Journal* 109 (1999): 165–247, CBA captures citizens' "desires" rather than preferences; Timur Kuran and Cass R. Sunstein, "Availability Cascades and Risk Regulation," *Stanford Law Review* 51 (1999): 683–768, CBA helps check irrational public risk perception; Mark Geistfeld, "Reconciling Cost-Benefit Analysis with the Principle that Safety Matters More than Money," *New York University Law Review* 76 (2001): 114–189. See generally Matthew D. Adler and Eric A. Posner, eds., *Cost-Benefit Analysis: Legal, Economic, and Philosophical Perspectives* (Chicago: University of Chicago Press, 2001).

16. Mark Sagoff, *The Economy of the Earth* (New York: Cambridge University Press, 1988). See also Cass R. Sunstein, "Endogenous Preferences, Environmental Law," *Journal of Legal Studies* 22 (1993): 217–254, 254; Guido Calabresi, "The Pointlessness of Pareto: Carrying Coase Further," *Yale Law Journal* 100 (1991): 1211–1237; Ronald Dworkin, "Is Wealth a Value?," *Journal of Legal Studies* 9 (1980): 191–226; Jules L. Coleman, *Markets, Morals and the Law* (New York: Cambridge University Press, 1988), 93 ("every economic notion of efficiency is of derivative and limited use in the public policy arena").

17. Committee on Risk Assessment of Hazardous Air Pollutants, National Research Council, *Science and Judgment in Risk Assessment* (Washington, D.C.: National Academy Press, 1994); National Research Council, *Risk Assessment in the Federal Government, Managing the Process* (Washington, D.C.: National Academy Press, 1983).

18. For a more detailed treatment of this issue, see Driesen, "The Societal Cost of Environmental Regulation," 605–612.

19. See Thomas O. McGarity, *Reinventing Rationality: The Role of Regulatory Analysis in the Federal Bureaucracy* (New York: Cambridge University Press, 1991), 131; Winston Harrington, Richard D. Morgenstern, and Peter Nelson, "On the Accuracy of Regulatory Cost Estimates," RFF Discussion Paper 99-18 (1999), 3, 5–7, reviewing the literature. Harrington, Morgenstern, and Nelson's review of a limited class of regulation confirms that ex ante overestimates of cost are common, but claim that underestimation exists in a few cases and accurate estimation is fairly common (23). They define accurate estimation, however, to allow for rather substantial deviations from the estimate (13).

20. McGarity, *Reinventing Rationality,* 131–132.

21. See, e.g., David Copp, "Morality, Reason, and Management Science," in *Ethics and Economics,* ed. Ellen Frankel Paul, Fred D. Miller Jr., Jeffrey Paul (New York: B. Blackwell, 1985), 131; Congressional Record, 104th Cong., 1st sess., 1995, 141, pt. 118:10392, statement of Senator Roth; Congressional Record, 104th Cong., 1st sess., 1995, 141, pt. 114:9995, statement of Senator Murkowski; Congressional Record, 104th Cong., 1st sess., 1995, 141, pt. 111:9674, statement of Senator Kyle.

22. Producers cannot pass production price increases on to consumers for highly price-elastic goods because charging more will simply cause consumers to substitute

other products. See generally John M. Gowdy and Sabine O'Hara, *Economic Theory for Environmentalists* (Delray Beach, Fla.: St. Lucie Press, 1995), 109.

23. Energy Information Administration, *Annual Energy Review 1999* (Washington, D.C.: U.S. Department of Energy, 2000), table 8.2, showing coal as providing more than 50% of electricity generation.

24. See Julie Edelson Halpert, "Harnessing the Sun and Selling it Abroad: U.S. Solar Industry in Export Boom," *New York Times,* 5 June 1996, D1, D20.

25. See Ralph C. Cavanagh, "Least-Cost Planning Imperatives for Electric Utilities and Their Regulators," *Harvard Environmental Law Review* 10 (1986): 299–344, 315.

26. See generally Barbara White, "Coase and the Courts: Economics for the Common Man," *Iowa Law Review* 72 (1987): 577–635, 593, arguing that since one cannot predict the implications of a transaction throughout the economy, one cannot predict whether a transaction justified by traditional economic efficiency tests is, in fact, economically efficient for the society.

27. See, e.g., Congressional Record, 104th Cong., 1st sess., 1995, 141, pt. 111:9650, statement of Senator Hutchison; Congressional Record, 103d Cong., 2d sess., 1994, 140, pt. 91:H5748–5749, statement of Rep. Delay. Cf. Congressional Record, 104th Cong., 1st sess., 1995, 141, pt. 116:10215, statement of Senator Biden, citing studies that show little correlation between regulation and lack of competitiveness.

28. See Adam B. Jaffee, Steven R. Peterson, Paul R. Portney, and Robert Stavins, *Environmental Regulation and International Competitiveness: What Does the Evidence Tell Us?* (Washington, D.C.: Resources for the Future, 1994); Stephen M. Meyer, *Environmentalism and Economic Prosperity: Testing the Environmental Impact Hypothesis* (1992)(unpublished manuscript on file with the author).

29. See Michael E. Porter and Claas Van der Linde, "Toward a New Conception of the Environment-Competitiveness Relationship," *Journal of Economic Perspectives* 9 (fall 1995): 97–118; Michael E. Porter, "America's Green Strategy," *Scientific American,* April 1991, 168; Michael E. Porter, *The Competitive Advantage of Nations* (New York: Free Press, 1990), 647–649.

30. See Porter and Van der Linde, "Environment-Competitiveness Relationship," 98.

31. See ibid. See generally Porter, *Competitive Advantage,* providing detailed case studies and more fully articulating Professor Porter's general theory.

32. See generally White, "Coase and the Courts," 594; "Economists have demonstrated rigorously that when constraints on efficiency . . . exist throughout the economy, applying policies to induce efficiency between some of the parties is not necessarily or even likely to be an economic improvement for society as a whole."

33. Donald T. Hornstein, "Lessons from Federal Pesticide Regulation on the Paradigms and Politics of Environmental Law Reform," *Yale Journal on Regulation* 10 (1993): 369–446, 422.

34. Ibid., 437–438.

35. The Rand Institute has estimated that the average damages for an asbestosis claim equal $54,000.00 excluding punitive damages. James S. Kakalik, Patricia A. Ebener, William L.F. Felstiner, Gus W. Haggstrom and Michael G. Shanley, *Variations in Asbestos Litigation Compensation and Expenses, Rand Institute for Civil Justice* (Rand, Santa Monica, 1984), 30 (hereafter *Rand Study*). By 1980, litigants had settled 30,000 asbestos related claims. *Dunn v. Hovic,* 1 F.3d 1371, 1393–94 (3d Cir. 1993) (Weis, J., dissenting) (citing *In re School Asbestos Litigation,* 789 F.2d 996, 1000 (3d Cir. 1986)). The Rand Study estimates that 76% of all asbestos-related claims involve asbestosis claims. Rand Study at v. Multiplying 30,000 (the number of settled asbestos related claims) by .76 (percentage involving asbestosis) by $54,000 (the cost of an average asbestosis claim) yields a total of 1.2 billion dollars.

36. In 1990, at least 90,000 cases were pending involving asbestos claims. See *Dunn v. Hovic,* 1 F.3d 1371, 1394 (3d Cir. 1993) (Weis J., dissenting), citing the Judicial Conference Ad Hoc Comm. on Asbestos Litigation, "Report to the Chief Justice of the United States and Members of the Judicial Conference of the United States" (1991), reprinted in *Asbestos Litigation Rep.*, March 1991 22698, 22702–30 for an estimate of 90,000 claims and citing a 160,000 claim estimate in State Judges Asbestos Litig. Comm., "Megatorts: The Lessons of Asbestos Litigation" (July 21, 1992), reprinted in *Mealey's Litig. Rep.—Asbestos,* November 20, 1992, at B–1). In addition, asbestos experts estimate that 668,363 new claims will be filed by 2049. *Dunn,* 1 F.3d 1394 (Weiss, J., dissenting) (citing Eric Stallard and Kenneth Manton, "Estimates and Projections of Asbestos-Related Mesothelioma and Exposures among Manville Personal Injury Settlement Trust Claimants, 1990-2049" (Draft Nov. 9, 1992), 42). Multiplying the 758,363 claims (90,000 pending plus 668,363 cases projected) by .76 (percentage of asbestos related claims likely to involve asbestosis, infra n. 18) by $54,000.00 (the approximate value of an asbestos claim, infra n. 18) yields $31.7 billion. This estimate probably understates the financial cost, since it excludes cases filed between 1987 and 1989, fails to account for inflation since 1982, and uses lower bound estimates for components in the equation.

37. *Corrosion Proof Fittings v. EPA,* 947 F.2d 1201, 1219 (5th Cir. 1991). For a critique of *Corrosion Proof Fittings,* see Thomas O. McGarity, "The Courts and the Ossification of Rulemaking: A Response to Professor Seidenfeld," *Texas Law Review* 75 (1997): 525–558, 541–549.

38. See, e.g., *Ethyl Corp. v. EPA,* 541 F.2d 1 (D. C. Cir. 1976).

39. *Corrosion Proof Fittings,* 947 F.2d 1214.

40. See *Industrial Union Dept., AFL-CIO v. American Petrol. Inst.,* 448 U.S. 607 (1980).

41. *Baltimore Gas and Electric Co. v. Natural Resources Defense Council,* 462 U.S. 87, 103 (1983), holding that a court must be "at its most deferential" when reviewing scientific determinations citing dissenting and concurring opinions in *Benzene.* See also *Commission v. Florida Power and Light Co.,* 404 U.S. 453 (1972), applying a more deferential approach to agency scientific decisions.

42. See, e.g., *Corrosion Proof Fittings,* 947 F.2d 1214; *International Union v. OSHA,* 938 F.2d 1310, 1322 (D.C. Cir. 1991).

43. The D.C. Circuit, which has jurisdiction over national rules under most federal environmental statutes, apparently considers the arbitrary and capricious standard to require substantial evidence. See *Association of Data Processing Service Organizations, Inc. v. Board of Governors of the Fed. Reserve System,* 745 F.2d 677, 683 (D.C. Cir. 1984); *Associated Industries v. U.S. Dept. of Labor,* 487 F.2d 342 (2nd Cir. 1973); *Consumers Union of U.S. Inc. v. FTC,* 801 F.2d 417, 422 (D.C.Cir. 1986). Cf. *Corrosion Proof Fittings,* 947 F.2d 1213, concluding that the substantial evidence standard is different from the arbitrary and capricious standard.

44. Hornstein, "Lessons from Federal Pesticide Regulation on the Paradigms and Politics of Environmental Law Reform," 436–437 and n. 395, explaining why risk assessment creates "strategic incentives to avoid a serious scientific examination of "true" levels of public health and environmental risk" and detailing falsification.

45. See David M. Driesen, "Five Lessons from the Clean Air Act Implementation," *Pace Environmental Law Review* 14 (1996): 51–62.

Chapter 3

1. The definition of free trade, however, is not entirely clear. See David M. Driesen, "What Is Free Trade?: The Real Issue Lurking Behind the Trade and Environment Debate," *Virginia Journal of International Law* 41 (2001): 279–368.

2. See Edith Brown Weiss, "Understanding Compliance with International Environmental Agreements: The Baker's Dozen Myths," *University of Richmond Law Review* 32 (1999): 1555–1589, 1555, more than 1000 international legal instruments.

3. See Chris Wold, "Multilateral Environmental Agreements and the GATT: Conflict and Resolution?," *Environmental Law* 26 (1996): 841–921, 870–874.

4. Michael J. Glennon, "Has International Law Failed the Elephant?," *American Journal of International Law* 84 (1990): 1–43, 25–26, ban on trade in leopard parts have been far more successful then a ban on trade in rhinoceros horns.

5. See Elizabeth P. Barratt-Brown, "Building a Monitoring and Compliance Regime Under the Montreal Protocol," *Yale Journal of International Law* 16 (1991): 519–570.

6. "Montreal Treaty Seen as Major Success in Effort to Protect Stratospheric Ozone Layer," *Environmental Reporter* (BNA) 28 (August 29, 1997): 778.

7. See generally Richard Elliot Benedick, *Ozone Diplomacy: New Directions in Safeguarding the Planet* (Cambridge, Mass.: Harvard University Press, 1991).

8. See Steve Charnovitz, "Environmental Trade Sanctions and the GATT: An Analysis of the Pelly Amendment on Foreign Environmental Practices," *American University Journal of International Law and Policy* 9 (1994): 751–807.

9. Steve Charnovitz, "Free Trade, Fair Trade, Green Trade: Defogging the Debate," *Cornell International Law Journal* 27 (1994): 459–525, 493.

10. See David M. Driesen, "The Congressional Role in International Environmental Law and its Implications for Statutory Interpretation," *Boston College Environmental Affairs Law Review* 19 (1991): 287–315, 303–305.

11. See Alan O. Sykes, "Comparative Advantage and the Normative Economics of International Trade Policy," *Journal of International Economic Law* 1 (1998): 49–82.

12. See Adam Smith, *An Inquiry into the Nature and Causes of the Wealth of Nations*, ed. Edwin Cannan (New York: Modern Library, 1776).

13. See David Ricardo, "On the Principles of Political Economy and Taxation," in vol. 1 of *The Works and Correspondence of David Ricardo* (Cambridge: Cambridge University Press, 1962).

14. General Agreement on Tariffs and Trade, Oct. 30, 1947, 61 Stat. A-11, T.I.A.S. No. 1700, 55 U.N.T.S. 194 [hereafter GATT].

15. Ibid., art. III.

16. Ibid., art. XI.

17. Ibid., art. I.

18. Ibid., art. XX.

19. Ibid., art. XX (chapeau).

20. Benedick, *Ozone Diplomacy,* 91.

21. See Dricsen, "What Is Free Trade?," 338–340; Steve Charnovitz, "Exploring the Environmental Exceptions in GATT Article XX," *Journal of World Trade* vol. 25, no. 5 (1991): 37–55.

22. 1994, Final Act Embodying the Results of the Uruguay Round of Multinational Trade Negotiations, Annex 2, arts. 16(4), 17(14), reprinted in *The Results of the Uruguay Round of Multilateral Trade Negotiations: The Legal Texts* (Geneva: GATT Secretariat, 1994), 417, 419.

23. "WTO Report of the Appellate Body on U.S. Import Prohibitions of Certain Shrimp and Shrimp Products," *I.L.M.* 38 (October 12, 1998): 118, 127–134.

24. For example, the tuna/dolphin decisions treated measures that demanded process changes in imported goods as a condition of imports as quantitative restrictions, prohibited by GATT article XI. Logically, a national restriction implementing an international agreement would also violate article XI, to the extent it required a process change as a condition upon importation. It is not clear that GATT would go this far, since these same opinions emphasized the evils of unilateralism, not multilateralism. But the uncertainty has influenced international environmental agreements.

25. See Driesen, "What Is Free Trade?," 300–301.

26. Agreement on the Application of Sanitary and Phytosanitary Measures, April 15, 1994, WTO Agreement, Annex 1A, reprinted in *GATT Secretariat, The Results*

of the Uruguay Round of Multilateral Trade Negotiations: The Legal Text (1994) (Geneva: GATT Secretariat, 1994) [hereafter SPS Agreement].

27. See WTO Dispute Settlement Panel Report on E.C.—Measures Concerning Meat And Meat Products (Hormones), 1997 WL 569984, 8.36–8.41.

28. See Steve Charnovitz, "The World Trade Organization, Meat Hormones, and Food Safety," *International Trade Reporter* (BNA) 14, no. 41 (1997): 1781–1787.

29. Report of the Appellate Body: EC Measures Concerning Meat and Meat Products (Hormones), 1998 WL 25520 186.

30. Ibid., 193.

31. Ibid., 194.

32. Ibid., 198.

33. Ibid., 196, 199–200.

34. Ibid., 207.

35. Ibid., 187.

36. See, e.g., Australia-Measures Affecting Importation of Salmon, AB-1998-5, October 20, 1998, reversing Panel assumption that a document containing "some evaluation" of risk is a risk assessment under the SPS agreement.

37. See Thomas O. McGarity, "Substantive and Procedural Discretion in the Administrative Resolution of Science Policy Questions: Regulating Carcinogens in EPA and OSHA," *Georgetown Law Journal* 67 (1979): 729–810, no regulatory program is possible if it must be based solely upon accepted "facts"; David A. Wirth, "International Trade Agreements: Vehicles for Regulatory Reform," *University of Chicago Legal Forum* 1997 (1997): 331–373, 343.

38. See Driesen, "What Is Free Trade?," 298.

39. See Vern R. Walker, "Keeping the WTO from Becoming the 'World Trans-science Organization': Scientific Uncertainty, Science Policy, and Fact-finding in the Growth Hormones Dispute," *Cornell International Law Journal* 31 (1998): 251–320, 318, Appellate Body may have required evidence "sufficiently specific and probative . . . to overcome a presumption of no risk."

40. See ibid; Appellate Body may have placed a "substantial burden indeed" upon defending countries; David A. Wirth, "The Role of Science in the Uruguay Round and NAFTA Trade Disciplines," *Cornell International Law Journal* 27 (1994): 817–859, 833, science can inform the regulatory process, but cannot determine results with particularity; Donna Roberts, "Preliminary Assessment of the Effects of the WTO Agreement on Sanitary and Phytosanitary Trade Regulations," *Journal of International Economic Law* 1 (1998): 377–405, 396–397. Roberts, however, believes that fears that SPS disciplines would occasion an "intolerable assault on . . . food safety and environmental standards have likely been overdrawn" (399). She cites the fact that most of the disputes to date involve developed countries as support for this view (398–399). She does not explain why SPS disciplines might not undermine food safety and environmental standards in developed countries.

Prior to the Beef Hormone decision a number of scholars believed that the SPS agreement did not interfere with standard setting. Obviously, the Beef Hormone decision requires reevaluation of these views.

41. SPS Agreement, art. 5.6.

42. See Driesen, "What Is Free Trade," 359–360.

43. See Alan O. Sykes, "Regulatory Protectionism and the Law of International Trade," *University of Chicago Law Review* 66 (1999): 1–46.

44. See generally, Charnovitz, "Free Trade, Fair Trade, Green Trade," 478, as the interdependence of economies increase, more environmental measures come within GATT's purview.

45. See Michael E. Porter and Claas Van der Linde, "Toward a New Conception of the Environment-Competitiveness Relationship," *Journal of Economic Perspectives* 9 (fall 1995): 97–118; Michael E. Porter, "America's Green Strategy," *Scientific American* (April 1991): 168; Michael E. Porter, *The Competitive Advantage of Nations* (New York: Free Press, 1990).

46. See Driesen, "What Is Free Trade?"

47. See Herman E. Daly, *Beyond Growth* (Boston: Beacon Press, 1996).

48. See Edith Brown Weiss, *In Fairness to Future Generations: International Law, Common Patrimony, and Intergenerational Equity* (Dobbs Ferry, N.Y.: Transnational Publishers, 1989).

49. Alternatives to Daly's views of sustainability exist, of course. See Markku Ollikainen, "Sustainable Development from the Viewpoint of Ethics and Economics," and Friedrich Hinterberger and Eberhard K. Seifert, "Reducing Material Throughput: A Contribution to the Measurement of Dematerialization and Sustainable Human Development," in *Environment, Technology, and Economic Growth,* Andrew Tylecote and Jan van der Staaten, eds. (Northampton, Mass.: E. Elgar, 1997), 39–54, 75–78.

50. See Herman E. Daly, "Sustainable Growth: An Impossibility Theorem," in *Valuing the Earth,* ed. Herman E. Daly and Kenneth N. Townsend (Cambridge, Mass.: MIT Press, 1992), 271.

51. Ibid.

52. See, e.g., Geoffrey M. Heal, *Valuing the Future: Economic Theory and Sustainability* (New York: Columbia University Press, 1998).

53. See WTO Appellate Body Report on U.S.-Import Prohibition of Certain Shrimp and Shrimp Products, *I.L.M.* 38 (October 12, 1998): 118, 154–157 (1999).

54. Ibid.

55. Kenneth Arrow, Bert Bolin, Roert Costanza, Partha Dasgupta, Carl Folke, C.S. Holling, Bengt-Owe Jansson, Simon Levin, Karl-Goran Maler, Charles Perrings, and David Pimentel, "Economic Growth, Carrying Capacity, and the Environment," *Science* 268 (April 1995): 520–521.

56. See generally Driesen, "What Is Free Trade?"

57. See, e.g., Heal, *Valuing the Future.*

58. For an interesting critique of law and economics based primarily on Daly's ideas of sustainable development see Douglas A. Kysar, "Sustainability and the Macroeconomic Analysis of Law," *Boston College Law Review* 43 (2001): 1–71.

Chapter 4

Much of this chapter is based upon David M. Driesen, "Is Emissions Trading an Economic Incentive Program?: Replacing the Command and Control/Economic Incentive Dichotomy," *Washington and Lee Law Review* 55 (1998): 289–350, and, to a lesser extent, David M. Driesen, "Free Lunch or Cheap Fix?: The Emissions Trading Idea and the Climate Change Convention," *Boston College Environmental Affairs Law Review* 26 (1998): 1–87 [hereafter Driesen, "Free Lunch"]. See also David M. Driesen, "Choosing Environmental Instruments in a Transnational Context," *Ecology Law Quarterly* 27 (2000): 1–52.

1. See *United States v. Ethyl Corp.,* 761 F.2d 1153, 1157 (5th Cir. 1985). The technology-based standards for air emissions from new automobiles are also performance standards. They dictate a precise limitation, not a precise method for achieving the limitation. See Clean Air Act, U.S. Code, vol. 42, sec. 7521(g)(1994). The technology-based regulations that states with dirty air must promulgate may also be performance standards. The Clean Air Act requires state plans to "provide for the implementation of all reasonably available control measures," but not through "command and control" mandates. Clean Air Act, U.S. Code, vol. 42, sec. 7502(c)(1)(1994). Rather, it requires "reductions in emissions from existing sources . . . as may be obtained through the adoption, at a minimum, of reasonably available control technology." Ibid. EPA and the courts have interpreted this statutory language as authorizing promulgation of numerical emission limitations that do not dictate the precise compliance method. See *Michigan v. Thomas,* 805 F.2d 176, 184–85 (6th Cir. 1986).

2. See Robert W. Hahn and Robert N. Stavins, "Incentive-Based Environmental Regulation: A New Era from an Old Idea?," *Ecology Law Quarterly* 18 (1991): 1–42, 5–6; "A performance standard typically identifies a specific goal . . . and gives firms some latitude in meeting this target. These standards do not specify the means, and therefore, provide greater flexibility. . . ."; Richard B. Stewart, "Regulation, Innovation, and Administrative Law: A Conceptual Framework," *California Law Review* 69 (1981): 1256–1337, 1268 ("Performance standards allow regulated firms flexibility to select the least costly or least burdensome means of achieving compliance.") Cf. Richard B. Stewart, "Controlling Environmental Risks through Economic Incentives," *Columbia Journal of Environmental Law* 13 (1988): 153–169, 158; "Regulatory commands dictate specific behavior by each plant, facility, or product manufacturer. . . ."

3. Louis G. Tornatzky and Mitchell Fleischer et al., *The Processes of Technological Innovation* (Lexington, Mass.: Lexington Books, 1990), 101.

4. See *American Petroleum Inst. v. EPA,* 52 F.3d 1113, 1119 (D.C. Cir. 1995), stating that EPA may not limit use of ethanol in reformulated gasoline because Clean Air Act mandates performance standards; *PPG Ind., Inc. v. Harrison,* 660 F.2d 628, 636 (5th Cir. 1981), stating that authority to set performance standards does not include authority to specify fuels.

5. See Clean Air Act, U.S. Code, vol. 42, secs. 7412(h)(1),(2); 7411(h)(1),(2) (1994).

6. See Clean Air Act, U.S. Code, vol. 42, secs. 7412(h)(3); 7411(h)(3) (1994).

7. See, e.g., Federal Water Pollution Control Act, U.S. Code, vol. 33, sec. 1314(b)(1)(A),(b)(2)(A) (1994); *E.I. du Pont de Nemours v. Train,* 430 U.S. 112, 122 and n.9 (1977); *American Petroleum Inst. v. EPA,* 787 F.2d 965, 972 (5th Cir. 1986); *Association of Pac. Fisheries v. EPA,* 615 F.2d 794, 802 (9th Cir. 1980); *American Paper Inst. v. Train,* 543 F.2d 328, 340–42 (D.C. Cir. 1976); *American Iron and Steel Inst. v. EPA,* 526 F.2d 1027, 1045 (3d. Cir. 1975), modified, 560 F.2d 589 (3d. Cir. 1977).

8. Compare Hahn and Stavins, "Incentive-Based Environmental Regulation," 5, with Bruce A. Ackerman and William T. Hassler, "Beyond the New Deal: Coal and the Clean Air Act," *Yale Law Journal* 89 (1980): 1466–1571, 1481–1488, discussing NSPS that allegedly mandated flue gas scrubbing; Bruce A. Ackerman and William T. Hassler, *Clean Coal/Dirty Air* (New Haven, Conn.: Yale University Press, 1981), 15–21.

9. *Sierra Club v. Costle,* 657 F.2d 298, 312 (D.C. Cir. 1981).

10. Ibid., 324, 327–328, 340–343, 346–347.

11. See ibid., 368–373; Ackerman and Hassler, "New Deal," 1481; Bruce A. Ackerman and William T. Hassler, "Beyond the New Deal: Reply," *Yale Law Journal* 90 (1981): 1412–1434, 1421–1422, n. 43. Cf. Howard Latin, "Ideal Versus Real Regulatory Efficiency: Implementation of Uniform Standards and 'Fine-Tuning' Regulatory Reforms," *Stanford Law Review* 37 (1985): 1267–1332, 1277 n. 41, noting that standard allows using coal washing as offset, decreasing the percentage reduction needed from scrubbing; Ackerman and Hassler, *Clean Coal,* 15, 66–68, noting that coal washing reduces any given emissions base by only 20–40 percent, but replacing new source standards with less stringent reduction requirement that also applies to existing sources would produce better results.

12. Stewart, "Regulation, Innovation, and Administrative Law," 1269.

13. See Kurst Strasser, "Cleaner Technology, Pollution Prevention, and Environmental Regulation," *Fordham Environmental Law Journal* 9 (1997): 1–106, 32, innovation sometimes results from emission and discharge limits. See, e.g., U.S. Congress, Office of Technology Assessment, *Gauging Control Technology and Regulatory Impacts in Occupational Safety and Health—An Appraisal of OSHA's Analytical Approach,* OTA-ENV-635, 64 (Washington D.C.: U.S. Government Printing Office, 1995) [hereafter OTA Study]; Nicholas A. Ashford and George R. Heaton Jr., "Regulation and Technological Innovation in the Chemical Industry," *Law and Contemporary Problems* 46, no. 3 (1983): 109–157, 139–140.

14. OTA Study, 89. Nicholas A. Ashford, Christine Ayers, Robert F. Stone, "Using Regulation to Change the Market for Innovation," *Harvard Environmental Law Review* 9 (1985): 419–466, 440–441.

15. OTA Study, 90.

16. Ibid., 95. OSHA anticipated this possibility, but not the extent to which it dominated compliance strategies.

17. See Strasser, "Cleaner Technology, Pollution Prevention, and Environmental Regulation," 28–29.

18. See Benefits of the CFC Phaseout (visited January 24, 2001) <http://www.epa.gov/ozone/geninfo/benefits.html> (citing "aqueous cleaning" as an example of a cleaning process that reduced cost in phasing out CFCs); ICOLP Technical Committee, Eliminating CFC-113 and Methyl Chloroform in Precision Cleaning Operations (Washington, D.C.: United States Environmental Protection Agency, 1994), 114 (defining "aqueous cleaning" as cleaning parts with water to which suitable detergents, saponifers or other additives may be added).

19. See Ashford, Ayers, and Stone, 437, describing separation of process from cooling water to reduce contact with mercury as a "significant process innovation."

20. Cf. Daniel H. Cole and Peter Z. Grossman, "When Is Command-and-Control Efficient? Institutions, Technology, and the Comparative Efficiency of Alternative Regulatory Regimes for Environmental Protection," *Wisconsin Law Review* 1999: 887–938, 892, "empirical studies" do not demonstrate that "command-and-control" regulation is "invariably less efficient than" emissions trading.

21. See, e.g., J. H. Dales, *Pollution, Property, and Prices* (Toronto: University of Toronto Press, 1980), 92–100.

22. See James T. B. Tripp and Daniel J. Dudek, "Institutional Guidelines for Designing Successful Transferable Rights Programs," *Yale Journal on Regulation* 6 (1989): 369–391, 374.

23. $40,000 + $40,000= $80,000.

24. $40,000 + $120,000 = $160,000.

25. See generally Kenneth J. Arrow, *Social Choice and Individual Values* (New Haven, Conn.: Yale University Press, 1963); James M. Buchanan and Gordon Tullock, *The Calculus of Consent: Logical Foundations of Constitutional Democracy* (Ann Arbor: University of Michigan Press, 1962); Morris P. Fiorina, *Congress: Keystone of the Washington Establishment* (New Haven, Conn.: Yale University Press, 1989); David R. Mayhew, *Congress: The Electoral Connection* (New Haven, Conn.: Yale University Press, 1974); Dennis G. Mueller, Public Choice (New York: Cambridge University Press, 1979); William H. Riker, *Liberalism against Populism: A Confrontation between the Theory of Democracy and the Theory of Social Choice* (San Francisco: W. H. Freeman, 1982); Armatya K. Sen, *Collective Choice and Social Welfare* (San Francisco: Holden-Day, 1970); Daniel A. Farber and Philip E. Frickey, "The Jurisprudence of Public Choice," *Texas Law Review* 65 (1987): 873–927; Mark Kelman, "On Democracy-Bashing: A Skeptical

Look at the Theoretical and 'Empirical' Practice of the Public Choice Movement," *Virginia Law Review* 74 (1988): 199–273.

26. See James M. Buchanan and Gordon Tullock, "Polluters' Profits and Political Response: Direct Control Versus Taxes," *American Economic Review* 65 (1975): 139–147, 141–142.

27. See Hahn and Stavins, "Incentive-Based Environmental Regulation," 8–9 and n. 33, emissions trading tends to reach equilibrium.

28. See Tripp and Dudek, "Institutional Guidelines for Designing Successful Transferable Rights Programs," 374.

29. Ibid.

30. See Byron Swift, "The Acid Rain Test," *Environmental Forum* 14, no. 3 (May–June 1997): 16–25, 18, explaining that emission rates do not necessarily prevent increases in mass of emissions. Traditional regulations can limit pollution by mass rather than by rate. Hence, traditional regulation and emissions trading based on rates fail to constrain emissions in the face of growth in production, but limits on mass, whether expressed in performance standards or tradable allowances, may constrain emissions in the face of growth.

31. See *Texas Mun. Power Agency v. EPA,* 89 F.3d 858, 861 (D.C. Cir. 1996), involving claim seeking additional emission allowances; *Indianapolis Power and Light Co. v. EPA,* 58 F.3d 643, 647 (D.C. Cir. 1995) (same); *Madison Gas and Elec. Co. v. EPA,* 25 F.3d 526 (7th Cir. 1994) (same); *Monongahela Power Co. v. Reilly,* 980 F.2d 272-274 (4th Cir. 1992) (same).

32. See David M. Driesen, "Five Lessons from Clean Air Act Implementation," *Pace Environmental Law Review* 14 (1997): 51–62, 53–55.

33. See Tripp and Dudek, "Institutional Guidelines for Designing Successful Transferable Rights Programs," 370–371, explaining that designing and implementing emissions trading is technically complex; see also Natural Resources Defense Council, "Comments on Economic Incentive Program Rules and Related Guidance 58 Federal Register 11,110 (1993)" (on file with author); *Texas Mun. Power Agency,* 89 F.3d 867–875, upholding EPA's use of State-wide average baseline emission rate as basis for calculating emission allowances, its analysis of that data, and its decision regarding adjustment of allowances to account for prolonged outages during baseline period; *Indianapolis Power and Light Company,* 58 F.3d 647, rejecting claims to adjustment of allowances based on power outages during period used to establish baseline emissions.

Some emissions trading proponents seek to minimize the importance of these issues by dismissing them as "transitional" issues. See Bruce A. Ackerman and Richard B. Stewart, "Reforming Environmental Law: The Democratic Case for Market Incentives," *Columbia Journal of Environmental Law* 13 (1988): 171–199, 185. Because most responsible proponents recognize that emissions trading cannot wholly supplant traditional regulation, issues regarding their interaction are likely to be with us for a long time. See Dales, *Pollution, Property, and Prices,* 98, contending that emissions trading is impracticable for "diffuse" pollution; William

J. Baumol and Wallace E. Oates, *The Theory of Environmental Policy,* 2d ed. (New York: Cambridge University Press, 1988), 190; "The ideal policy package contains a mixture of instruments, with taxes, marketable permits, [and] direct control . . . each used in certain circumstances to regulate the sources of environmental damage"; Hahn and Stavins, "Incentive-Based Environmental Regulation," 15; "The best set of policies will typically involve a mix of market and more conventional regulatory processes". Professors Ackerman and Stewart do not explain why the issues complicating emissions trading would disappear over time. While these issues might be resolved appropriately given sufficient political will, it's hard to imagine their disappearance.

34. See *American Mun. Power-Ohio v. EPA,* 98 F.3d 1372, 1374–75 (D.C.Cir. 1996), rejecting creation of shutdown credits that would impede realization of pollution reductions; Jo Anne H. Aplet, *NOX/SOX Reclaim Implementation* (Los Angeles: American Lung Association, 1995), 6–7, 12–13, noting that shutdown credits are used under California rules.

35. The Kyoto Protocol, as interpreted at the Sixth Conference of the Parties at Bonn, does authorize credits for some types of forestry projects. See Conference of the Parties, Sixth Sess., part 2, Decision 5/CP.6, Annex at 7–9, FCCC/CP/2001/L.7 (July 24, 2001) (accepting afforestation and reforestation projects in developing countries as credit generating mechanisms).

36. Hahn and Stavins, "Incentive-Based Environmental Regulation," 8, n.33.

37. Ibid., 8–9, n.33.

38. See David A. Malueg, "Emissions Credit Trading and the Incentive to Adopt New Pollution Abatement Technology," *Journal of Environmental Economics and Management* 16 (1989): 52–57, 54; David Wallace, *Environmental Policy and Industrial Innovation: Strategies in Europe, The U.S. and Japan* (Washington, D.C.: Earthscan Publications, 1995), 20, explaining that Malueg's "more sophisticated model" casts doubt on the claim that emissions trading necessarily spurs innovation.

39. See Driesen, "Free Lunch."

40. See Richard A. Liroff, *Reforming Air Pollution Regulation: The Toil and Trouble of EPA's Bubble* (Washington, D.C.: Conservation Foundation, 1986), 100; "Most innovations under bubbles merely are rearrangements of conventional technologies."

41. See, e.g., Byron Swift, "Command without Control: Why Cap-and-Trade Should Replace Rate Standards for Regional Pollutants," *Environmental Law Reporter* 31 (2001): 10330–10341.

42. For accounts of the program, see Suzi Kerr and David Mare, "Market Efficiency in Tradeable Permit Markets with Transaction Costs: Empirical Evidence from the United States Lead Phasedown" in Suzi Clare Kerr, *Contracts and Tradeable Permit Markets in International and Domestic Environmental Protection* (Ph. D. dissertation, Harvard University, 1995); Robert W. Hahn and Gordon L. Hester, "Marketable Permits: Lessons for Theory and Practice," *Ecology Law Quarterly* 16 (1989): 361–406, 380–391. I refer to this as an example of trading for simplicity's

sake. Because the rule authorized intertemporal trades, this rule also exemplifies "banking" of emission credits.

43. Environmental Protection Agency, Final Rule, "Regulation of Fuels and Fuel Additives; Banking of Lead Rights," *Federal Register* 50, no. 10 (2 April 1985): 13116.

44. Ibid., 13177, 13127 (codified at EPA Air Programs, 40 C.F.R. sec. 80.20(e)(2)(1988)).

45. See ibid., 13119; Hahn and Hester, "Marketable Permits," 382, n. 125; U.S. General Accounting Office, *Vehicle Emissions: EPA Program to Assist Leaded-Gasoline Producers* (Washington, D.C.: General Accounting Office, 1986), 20.

46. See GAO, *Vehicle Emissions,* 3–4, 18–19, 23–24, citing failure to enforce against 25 potential violators, 49 cases of claimed credits not matching claimed sales of credits, error rates in reporting between 14% and 49.2% and no verification of compliance. Cf. Hahn and Hester, "Lessons," 388, n. 146.

47. See, e.g., Swift, "Command without Control."

48. Ibid., 10331.

49. Swift does claim that trading was essential to two technologies. Swift, "Command without Control," 10338. One of those "technologies," trading, is a transaction, not a technology. He does not claim that the other technology, power shifting, is an innovation. Indeed, the shifting of dispatch orders to use cleaner units more intensively than dirty units is a well-understood operational option.

50. See David Popp, "Pollution Control Innovations and the Clean Air Act of 1990," Working Paper #8593, NBER, 2001.

51. See A. Denny Ellerman et al., *Markets for Clean Air: The U.S. Acid Rain Program* (New York: Cambridge University Press, 2000), 130.

52. See Driesen, "Free Lunch."

53. See Frederick R. Anderson et al., *Environmental Improvement through Economic Incentives* (Baltimore, Md.: Johns Hopkins University Press, 1977), 15.

54. See ibid., 7; "As a practical matter total social damages are almost impossible to compute."

55. See Baumol and Oates, *Theory of Environmental Policy,* 178; Howard Gensler, "The Economics of Pollution Taxes," *Journal of Natural Resources and Environment Law* 10 (1994–1995): 1–12, 10–12.

56. See Mikael Skou Andersen, *Governance by Green Taxes: Making Pollution Prevention Pay* (New York: St. Martin's Press, 1994), 27.

57. See, e.g., Thomas H. Tietenberg, *Environmental and Natural Resources Economics,* 3d ed. (Reading, Mass.: Addison-Wesley, 1992), 30; James R. Kahn, *The Economic Approach to Environmental and Natural Resources,* 2d ed. (Fort Worth: Harcourt College Publishers, 1997), 37.

58. See Sidney A. Shapiro and Robert L. Glicksman, "Goals, Instruments, and Environmental Policy Choice," *Duke Environmental Law and Policy Forum* 10 (2000): 297–325, 299–304.

59. Adam B. Jaffe, Richard G. Newell, and Robert N. Stavins, "Technological Change and the Environment" (*Resources for the Future Discussion Paper* 00–47 2000), 49.

Chapter 5

1. See generally David Strong, *Crazy Mountains: Learning from Wilderness to Weigh Technology* (Albany, N.Y.: State University of New York Press, 1995).

2. Adam B. Jaffe, Richard G. Newell, and Robert N. Stavins, "Technological Change and the Environment," *Resources for the Future Discussion Paper* 00–47 (2000), 4.

3. Jaffe, "Technological Change and the Environment," 8. Richard Stewart defines innovation as combining invention and diffusion, see Richard B. Stewart, "Regulation, Innovation, and Administrative Law: A Conceptual Framework," *California Law Review* 69 (1981): 1256–1377, 1282, but Jaffe, Newell, and Stavins, on the other hand, consider innovation and diffusion to be, in principle, two different things, that are difficult to distinguish economically. See Jaffe et al., "Technological Change and the Environment." This simply highlights the lack of agreement about the precise definition of innovation mentioned at the outset.

4. See Stewart, "Regulation, Innovation, and Administrative Law," 1261, 1279.

5. For a very generalized philosophical and skeptical view of technology's value, see Strong, *Crazy Mountains*.

6. See ibid., 78–79.

7. See ibid., 87–91. See generally John Kenneth Galbraith, *The Affluent Society* (Boston: Houghton Mifflin, 1960).

8. Sylvie Faucheux, Isabelle Nicolai, and Martin O'Connor, "Globalization, Competitiveness, Governance, and Environment: What Prospects for a Sustainable Development," in *Sustainability and Firms,* ed. S. Faucheux et al. (Northampton, Mass: E. Elgar, 1998), 22.

9. See Robert Cowan and Staffan Hulten, "Escaping Technological Lock-in: The Case of the Electric Vehicle," *Technological Forecasting and Social Change* 53, no. 1 (1996): 61–79, 69.

10. See Jaffe et al., "Technological Change and the Environment," 11.

11. See Arnold W. Reitze, Jr., "Mobile Source Air Pollution Control," *Environmental Lawyer* 6 (2000): 309–439, 317, explaining that increased use of SUVs and light-duty trucks reduced fuel economy and raised emissions.

12. See Bjarne S. Jensen and Kar-Yiu Wong, *Dynamics, Economic Growth, and International Trade* (Ann Arbor, Mich.: University of Michigan Press, 1997), 29.

13. See Mark H. Dorfman, Warren R. Muir, and Catherine G. Miller, *Environmental Dividends Cutting More Chemical Wastes* (New York: Inform, 1992), 44–46, explaining the role of regulation and high waste disposal costs in motivating source reduction activities.

14. Jaffe et al., "Technological Change and the Environment," 11.

15. See Jerry L. Mashaw and David L. Harfst, *The Struggle for Auto Safety* (Cambridge, Mass.: Harvard University Press, 1990), 2.

16. Existing coal-fired power plants account for 60% of national sulfur dioxide emissions, 25% of nitrogen oxides, a third of thirteen other priority hazardous air pollutants, and 32% of carbon dioxide in the United States. See Environmental Law Institute, *Cleaner Power: The Benefits and Costs of Moving from Coal Generation to Modern Power Technologies* (Washington, D.C.: Environmental Law Institute, 2001), 2.

17. See Natural Resources Defense Council and Public Service Electric and Gas Company, *Benchmarking Air Emissions of Electric Utility Generators in the U.S.* (1998), 38; EPA, *Latest Findings on National Air Quality: 1999 Status and Trends* (Washington, D.C.: Environmental Protection Agency, 2000), 12.

18. See Byron Swift, "Command without Control: Why Cap-and-Trade Should Replace Rate Standards for Regional Pollutants," *Environmental Law Reporter* 31 (2001): 10330–10341, reviewing federal utility regulation in detail.

19. Reitze, "Mobile Source Air Pollution Control."

20. See Roger Dower et al., *Frontiers of Sustainability: Environmentally Sound Agriculture, Forestry, Transportation, and Power Production* (Washington, D.C.: Island Press, 1997), 301.

21. See, e.g., Jonathan B. Wiener, "Global Environmental Regulation: Instrument Choice in Legal Context," *Yale Law Journal* 108 (1999): 677–800, 718 n. 162, claiming that least-cost abatement will make the world better off even if it discourages innovation, thus implying that innovation has no value other than lowering cost. Cf. David M. Driesen, "Free Lunch or Cheap Fix?: The Emissions Trading Idea and the Climate Change Convention," *Boston College Environmental Affairs Law Review* 26 (1998): 1–87, discussing the qualitative and intertemporal knowledge spillover effects of innovation in the context of climate change policy.

22. Cf. Jaffe et al., "Technological Change and the Environment," 30–32, pointing out that innovation can lower the cost of abatement and thereby increase the amount of abatement deemed optimal.

23. Ibid., 40–43.

24. See, e.g., Cowan and Hulten, "Escaping Technological Lock-in"; Robin Cowan and David Kline, *The Implications of Potential "Lock-In" in Markets for Renewable Energy*, NREL/TP-460-22112 (1996).

25. See Bjorn Johnson, "Institutional Learning and Clean Growth," in *Environment, Technology and Economic Growth*, ed. Andrew Tylecote and Jan van der Staaten (Northampton, Mass.: E. Elgar, 1998), 93–111.

26. See E.O. Wilson, *The Diversity of Life* (New York: W. W. Norton, 1999); E.O. Wilson, *Biodiversity II: Understanding and Protecting Our Biological Resources* (Washington, D.C.: Joseph Henry Press, 1999).

27. See Henry W. McGee, Jr., and Kurt Zimmerman, "The Deforestation of the Brazilian Amazon: Law, Politics, and International Cooperation," *University of Miami Inter-American Law Review* 21 (1990): 513-550, 527.

Chapter 6

1. See Edwin Mansfield, *The Economics of Technological Change* (New York: W. W. Norton, 1968), 104–106, explaining that many innovations fail, but that they can generate substantial returns.

2. Indeed, some analysts argue that the role of continuous innovation is increased. See, e.g., Richard Florida and Martin Kenney. "The New Age of Capitalism: Innovation Mediated Production." *Futures* 25 (July/August 1993): 637–651.

3. See Edwin Mansfield et al., *The Production and Application of New Industrial Technology* (New York: Norton, 1977), 16, threat of new entrants has stimulated existing firms to innovate.

4. See Klaus Lindegaard, "Environmental Law, Environmental Globalization, and Sustainable Technology," in *Environment, Technology, and Economic Growth*, ed. Andrew Tylecote and Jan van der Straaten (Northampton, Mass.: E. Elgar, 1998), 133.

5. See E.N. Brandt, *Growth Company: Dow Chemical's First Century* (East Lansing: Michigan State University Press, 1997).

6. See, e.g., Homa Bahrami and Stuart Evans, "Flexible Recycling and High-Technology Entrepreneurship," in *Understanding Silicon Valley: The Autonomy of an Entrepreneurial Region*, ed. Martin Kenney (Stanford, Calif.: Stanford University Press, 2000), 177–178, emphasizing the crucial role learning through failure and through doing plays in Silicon Valley's entrepreneurial culture.

7. Adam B. Jaffe, Richard G. Newell, and Robert N. Stavins, "Technological Change and the Environment," *Resources for the Future Discussion Paper* 00-47 (2000), 12.

8. See Herbert A. Simon, *Administrative Behavior: A Study of Decision-Making Processes in Administrative Organization* (New York: Macmillan, 1947).

9. See Jaffe et al., "Technological Change and the Environment," 12.

10. Ibid., 9–10.

11. Ibid., 10.

12. Richard C. Dorf, ed., *The Technology Management Handbook* (Boca Raton, Fla.: CRC Press, 1999), 3–12.

13. Jack Manno, *Commoditization and Its Impact on Environment and Society* (Boca Raton, Fla.: Lewis Publishers, 2000).

14. See Leslie A. Davis, "North Pacific Pelagic Driftnetting: Untangling the High Seas Controversy," *Southern California Law Review* 64 (1991): 1057–1102, 1066–1073; Amy Blackwell, "The Humane Society and Italian Driftnetters:

Environmental Activists and Unilateral Action in International Environmental Law," *North Carolina Journal of International Law and Commercial Regulation* 23 (1998): 313–340, 318.

15. See Paul Hawken, Amory Lovins, and L. Hunter Lovins, *Natural Capitalism: Creating the Next Industrial Revolution* (Boston: Little, Brown, 1999), 5.

16. Ibid., 24.

17. See, e.g., Florida, Richard. "Lean and Green: The Move to Environmentally Conscious Manufacturing," *California Management Review* 39, no. 1 (fall 1996): 80–105; Hawken et al., *Natural Capitalism.*

18. For a discussion of the debate on this issue see Michael S. Common, *Sustainability and Policy: Limits to Economics* (New York: Cambridge University Press, 1995), 94–96.

19. See Suzanne Iudicello, Michael Weber, and Robert Wieland, *Fish, Markets, and Fishermen: The Economics of Overfishing* (Washington, D.C.: Island Press, 1999).

20. See generally Michael J. Glennon, "Has International Law Failed the Elephant?," *American Journal of International Law* 84 (1990): 1–43, 25–26.

21. Cf. Ralph C. Cavanagh, "Least-Cost Planning Imperatives for Electric Utilities and Their Regulators," *Harvard Environmental Law Review* 10 (1986): 299–344, 315.

22. See James Salzman, "Beyond the Smokestack: Environmental Protection in the Service Economy," *UCLA Law Review* 47 (1999): 411–489.

23. See Molly O'Meara, "Harnessing Information Technologies for the Environment," in Lester R. Brown et al., *State of the World 2000* (New York: Norton, 2000), 121, 127–130, discussing the negative environmental effects of computer manufacturing, use, and disposal.

24. See Manno, *Commoditization and Its Impact on Environment and Society.*

25. See David M. Driesen, "Is Emissions Trading an Economic Incentive Program?: Replacing the Command and Control/Economic Incentive Dichotomy," *Washington and Lee Law Review* 55 (1998): 289–350, 323.

26. See Nicholas A. Ashford, Christine Ayers, and Robert F. Stone, "Using Regulation to Change the Market for Innovation," *Harvard Environmental Law Review* 9 (1985): 419–466, 426.

27. See Richard B. Stewart, "Regulation, Innovation, and Administrative Law: A Conceptual Framework," *California Law Review* 69 (1981): 1256–1377, 1283.

28. Ibid.

29. See Martin Kennedy and Richard Florida, "Venture Capital in Silicon Valley," in *Understanding Silicon Valley, The Anatomy of an Entrepreneurial Region,* ed. Martin Kenney (Stanford, Calif.: Stanford University Press, 2000), 121–123, discussing various funding sources for venture capital.

30. Environmental Law Institute, *Barriers to Environmental Technology Innovations and Use* (Washington, D.C.: Environmental Law Institute, 1998), 10–11.

31. See, e.g., *Natural Resources Defense Council v. EPA,* 655 F.2d 318 (D.C. Cir. 1981); *International Harvester Company v. Ruckelshaus,* 478 F.2d 615 (D.C. Cir. 1973).

32. See *International Harvester,* 478 F.2d 615.

33. See Stewart, "Regulation, Innovation, and Administrative Law," 1282.

Chapter 7

1. See generally Herbert A. Simon, *Administrative Behavior: A Study of Decision-Making Processes in Administrative Organization* (New York: Macmillan, 1948).

2. See ibid., 79–109.

3. See Richard C. Dorf, ed., *The Technology Management Handbook* (Boca Raton, Fla.: CRC Press, 1999), 1–11, describing the components of a business plan.

4. Ibid., 16–21.

5. See generally Oliver E. Williamson, *The Mechanisms of Governance* (New York: Oxford University Press, 1996), 17, all firms are bureaucracies.

6. See Edwin Mansfield et al., *The Production and Application of New Industrial Technology* (New York: Norton, 1977), 8–11, discussing pervasive uncertainty in technological innovation.

7. See generally Friedrich A. Von Hayek, *The Constitution of Liberty* (Chicago: University of Chicago Press, 1960), 124.

8. See generally Clayton M. Christensen, *The Innovator's Dilemma: When New Technologies Cause Great Firms to Fail* (Boston: Harvard Business School Press, 1997).

9. See generally ibid.

10. See John T. Preston, "Technology Innovation and Environmental Progress," in *Thinking Ecologically: The Next Generation of Environmental Policy,* ed. Marian R. Chertow and Daniel C. Esty (New Haven, Conn.: Yale University Press, 1997), 140–146.

11. See David Wallace, *Environmental Policy and Industrial Innovation: Strategies in Europe, the USA, and Japan* (Washington, D.C.: Earthscan Publications, 1995), 1, 2, 22.

12. See, e.g., Barry Bozeman, *All Organizations Are Public: Bridging Public and Private Organizational Theories* (San Francisco: Jossey-Bass, 1987).

13. Congress may exercise control through mechanisms other than just legislative specificity. See William T. Gormley, *Taming the Bureaucracy: Muscles, Prayers, and Other Strategies* (Princeton, N.J.: Princeton University Press, 1989), 194–223.

14. See J. Clarence Davies and Jan Mazurek, *Pollution Control in the United States: Evaluating the System* (Washington, D.C.: Resources for the Future, 1998), 146–147, discussing the limited impact of regulation on plant closings and unemployment; Stephen M. Meyer "The Economic Impact of Environmental

Regulation." *Journal of Environmental Law and Practice* 3 (September/October 1995): 4–16. See also Adam B. Jaffe, Steven R. Peterson, Paul R. Portney, and Robert N. Stavins, "Environmental Regulation and the Competitiveness of U.S. Manufacturing," *Journal of Economic Literature* 33, no. 1 (March 1995): 132–163.

15. See, e.g., Theodore J. Lowi, *The End of Liberalism: The Second Republic of the United States* (New York: Norton, 1979). For a brief review of some of the political science literature, see David Epstein and Sharyn O'Halloran, "The Nondelegation Doctrine and the Separation of Powers: A Political Science Approach," *Cardozo Law Review* 20 (1999): 947–987. See also David Schoenbrod, *Power without Responsibility: How Congress Abuses the People through Delegation* (New Haven, Conn.: Yale University Press, 1993). Cf. Peter H. Schuck, "Delegation and Democracy: Comments on David Schoenbrod," *Cardozo Law Review* 20 (1999): 775–793.

16. See Federal Water Pollution Control Act, U.S. Code, vol. 33, secs. 1251(a)(1), 1311 (1994).

17. See generally James Q. Wilson, *Bureaucracy: What Government Agencies Do and Why They Do It* (New York: Basic Books, 1989), 243.

18. See, e.g., Thomas O. McGarity, "The Internal Structure of EPA Rulemaking," *Law and Contemporary Problems* 54 (1991): 57–111.

19. See Wesley A. Magat, Alan J. Krupnik, and Winston Harrington, *Rules in the Making: A Statistical Analysis of Regulatory Agency Behavior* (Washington, D.C.: Resources for the Future, 1986), 19–25. See also Jerry L. Mashaw and David L. Harfst, *The Struggle for Auto Safety* (Cambridge, Mass.: Harvard University Press, 1990), 71.

20. See Mashaw and Harfst, *The Struggle for Auto Safety,* 97–98.

21. Frank B. Cross, "Shattering the Fragile Case for Judicial Review of Rule-making," *Virginia Law Review* 85 (1999): 1243–1334, 1315.

22. See ibid., 1327; Martin Shapiro, "APA: Past, Present, and Future," *Virginia Law Review* 72 (1986): 447–492, 452–454.

23. See "Litigation: EPA Loses about Half of Cases Filed in D.C. Circuit in Recent Years, Study Says," *Environment Reporter* 31 (June 2, 2000): 1145, 1183–1187.

24. See Shep Melnick, *Regulation and the Courts* (Washington, D.C.: Brookings Institution, 1983).

25. See *Chevron v. Natural Resources Defense Council,* 467 U.S. 837 (1983).

26. See Peter H. Shuck and E. Donald Elliott, "To the Chevron Station: An Empirical Study of Federal Administrative Law," *Duke Law Journal* 1990 (1991): 984–1077, 1025; Thomas W. Merrill, "Judicial Deference to Executive Precedent," *Yale Law Journal* 101 (1992): 969–1033, 982, 984; "Chevron, State Farm, and EPA in the Courts of Appeals during the 1990s," *Environmental Law Reporter* 31 (April 2001): 10371.

27. See, e.g., *United States v. Mead Corp.,* 533 U.S. 218, 221, 226–227 (2001), declining to apply broad Chevron type deference and substituting an amorphous

test for when it will apply; *Whitman v. American Trucking Ass'ns,* 531 U.S. 457, 481–483 (2001), declaring agency interpretation of an ambiguous statute without the bounds of the ambiguity; *Solid Waste Agency of Northern Cook County v. Army Corp of Engineers,* 531 U.S. 159, 168–171 (2001), declining to acquiesce in an administration interpretation of the definition of navigable waters under the Federal Water Pollution Control Act, despite reenactment of relevant provision.

28. See Riley A. Dunlap, "Public Opinion and Environmental Policy," in *Environmental Politics and Policy: Theories and Evidence,* ed. James P. Lester (Durham, N.C.: Duke University Press, 1995), 64–114.

29. See James Q. Wilson, *Bureaucracy: What Government Agencies Do and Why They Do It* (New York: Basic Books, 1989), 115, government agencies cannot control how they allocate their resources.

30. See McGarity, "The Internal Structure of EPA Rulemaking," describing EPA's team-based approach.

31. Clayton Christensen points out that sometimes the innovation can outstrip demand, leading established firms to eschew desirable innovation. See Christensen, *The Innovator's Dilemma.*

32. See Thomas O. McGarity, "Some Thoughts on 'Deossifying' the Rulemaking Process," *Duke Law Journal* 41 (1992): 1385–1462. See, e.g., Jerry L. Mashaw and David L. Harfst, "Regulation and Legal Culture: The Case of Motor Vehicle Safety," *Yale Journal on Regulation* 4 (1987): 257–316; Stephen Breyer, "Judicial Review of Questions of Law and Policy," *Administrative Law Review* 38 (1986): 363–398, 383; William F. West, *Administrative Rulemaking: Politics and Processes* (Westport, Conn.: Greenwood Press, 1985), 188.

33. Nor have political scientists seriously contested it. For example, one of the leading critics of sharp distinctions between public and private organizations, Barry Bozeman, has acknowledged that "red tape" constitutes a significant problem for public bureaucracy, even though careful thinking about its sources shows that privatization may face similar problems. See Barry Bozeman, *Bureaucracy and Red Tape* (Upper Saddle River, N.J.: Prentice Hall, 2000). This book, however, concerns itself with more than just "red tape" as narrowly defined by Bozeman. He defines "red tape" as "rules, regulations, and procedures" that "do not advance" the rules' "legitimate purposes." Ibid., 12. By contrast, many of the procedures this book concerns itself with may advance legitimate purposes (although some may not).

34. William S. Jordan, III, "Ossification Revisited: Does Arbitrary and Capricious Review Significantly Interfere with Agency Ability to Achieve Regulatory Goals Through Informal Rulemaking?," *Northwestern University Law Review* 94 (2000): 393–449, disputing the ossification claim.

35. See David M. Driesen, "The Societal Cost of Environmental Regulation: Beyond Administrative Cost-Benefit Analysis," *Ecology Law Quarterly* 24 (1997): 545–617, 596–597, 602–603; Michael Francis Gaheen, *Cost-Benefit Analysis and the Regulatory Process: A Case Study of the Environmental Protection Agency's Asbestos Regulation* (Ph.D. dissertation, University of Maryland).

36. See Robert V. Percival, Alan S. Miller, Christopher H. Schroeder, and James P. Leape, *Environmental Regulation: Law, Science, and Policy,* 3d ed. (Gaithersburg, Md.: Aspen Law and Business, 2000), 425–427, a decade of delay in strengthening benzene standard caused scores of deaths, if OSHA's risk assessment is accurate.

37. See Walter A. Rosenbaum, "The Bureaucracy and Environmental Policy," in *Environmental Politics and Policy: Theory and Evidence,* ed. James P. Lester (Durham, N.C.: Duke University Press, 1995), 231, finding the "literature on environmental administration . . . pervasively pessimistic about bureaucratic performance. . . ."

38. See ibid., 223–226. Cf. ibid., 207–211, noting that the statutes were initially designed to prevent regulatory capture through "action forcing" mechanisms and more "scientific" regulation.

39. Cf. ibid., 220; few researchers expect traditional capture to occur.

40. See ibid., 220–226.

41. See Magat et al., *Rules in the Making,* 157.

42. See Philip J. Harter, "Assessing the Assessors: The Actual Performance of Negotiated Rulemaking," *New York University Environmental Law Journal* 9 (2000): 32–59. Cf. Cary Coglianese, "Assessing Consensus: The Promise and Performance of Negotiated Rulemaking," *Duke Law Journal* 46 (1997): 1255–1349; Charles C. Caldart and Nicholas A. Ashford, "Negotiation as a Means of Developing and Implementing Environmental and Occupational Health and Safety Policy," *Harvard Environmental Law Review* 23 (1999): 141–202.

43. See Elizabeth Cook, ed., *Ozone Depletion in the United States: Elements of Success* (Washington, D.C.: World Resources Institute, 1996).

44. See David Hunter, James Salzman, and Durwood Zaelke, *International Environmental Law and Policy* (New York: Foundation Press, 1998), 561.

45. See Greenpeace, *Money to Burn: The World Bank, Chemical Companies and Ozone Depletion* (Washington, D.C.: Greenpeace USA, 1994).

46. See ibid.

47. See David M. Driesen, "Choosing Environmental Instruments in a Transnational Context," *Ecology Law Quarterly* 27 (2000): 1–52, 6, 15–16.

48. See generally David G. Victor et al., eds., *The Implementation and Effectiveness of International Environmental Commitments: Theory and Practice* (Cambridge, Mass.: MIT Press, 1998), 4, noting the necessity for national action affecting private conduct in a book assessing such actions; Edith Brown Weiss and Harold K. Jacobson, eds., *Engaging Countries: Strengthening Compliance with International Environmental Accords* (Cambridge, Mass.: MIT Press, 1998), presenting case studies of national compliance efforts.

49. See David A. Wirth, "The International Trade Regime and the Municipal Law of Federal States: How Close a Fit?," *Washington and Lee Law Review* 49 (1992): 1389–1401, 1391, stating that the current international environmental system invites "holdouts, free riders, laggards, scofflaws, and defectors."

Chapter 8

1. See Marian R. Chertow and Daniel C. Esty, eds., *Thinking Ecologically: The Next Generation of Environmental Policy* (New Haven, Conn.: Yale University Press, 1997), 2.

2. See Lester R. Brown et al., *State of the World 2000* (New York: Norton, 2000), 5.

3. Ibid.

4. See David M. Driesen, "The Societal Cost of Environmental Protection: Beyond Administrative Cost-Benefit Analysis," *Ecology Law Quarterly* 24 (1997): 545–617, 583–584.

5. See OECD Environmental Data: Compendium 1999 (Paris: Organization for Economic Co-operation and Development).

6. See EPA, *National Air Quality and Emissions Trends Report, 1999* (Washington D.C.: United States Environmental Protection Agency, 2001), 17–20.

7. See EPA, *Latest Findings on National Air Quality: 1999 Status and Trends* (Research Triangle Park, N.C.: United States Environmental Protection Agency, Office of Air Quality Planning and Standards, 2000), 23; EPA Air Programs, C.F.R., vol. 40, secs. 82.6; 82.7; 82.270 (2000); "Montreal Treaty Seen as Major Success in Effort to Protect Stratospheric Ozone Layer," *Environment Reporter* 28 (August 29, 1997), 778.

8. See Lee Anne Duval (note), "The Future of the Montreal Protocol: Money and Methyl Bromide," *Virginia Environmental Law Journal* 18 (1999): 609–637, arguing that developed countries have failed to meet a legal obligation to fund developing country phase out of methyl bromide.

9. See "Environmental Investigating Group Finds Widespread Trade of CFCs in Europe," *International Environmental Reporter* 20 (September 17, 1997), 869.

10. See EPA, *Latest Findings on National Air Quality,* 23.

11. See National Research Council, *Rethinking the Ozone Problem in Urban and Regional Air Pollution* (Washington, D.C.: National Academy Press, 1991), 24–27.

12. See generally Environmental Protection Agency, Notice of Proposed Policy, "Interim Implementation Policy on New or Revised Ozone and Particulate Matter (PM) National Ambient Air Quality Standards (NAAQS)," *Federal Register* 61, no. 241 (December 13, 1996): 65752, 65756–57.

13. See *Whitman v. American Trucking Ass'ns,* 531 U.S. 457, 462–63 (2001).

14. See Arnold W. Reitze, Jr., "Mobile Source Air Pollution Control," *Environmental Lawyer* 6 (2000) 309–439.

15. See Emil Frankel, "Coexisting with the Car," in *Thinking Ecologically,* 190.

16. See Reitze, "Mobile Source Air Pollution Control," 316, 413.

17. See Robert W. Adler, Jessica C. Landman, and Diane M. Cameron, *The Clean Water Act 20 Years Later* (Washington, D.C.: Island Press, 1993), 14.

18. Ibid.

19. See ibid., 16, describing trends in regulation of point sources and remaining problems.

20. See Robert Adler, "Controlling Nonpoint Source Water Pollution: Is Help on Its Way (from the Courts or EPA)?," *Environmental Law Reporter* 31 (March 2001): 10270.

21. See Oliver A. Houck, "TMDLs: The Resurrection of Water Quality Standards-Based Regulation under the Clean Water Act," *Environmental Law Reporter* 27 (July 1997): 10329; Oliver A. Houck, "TMDLs, Are We There Yet?: The Long Road toward Water Quality-Based Regulation Under the Clean Water Act," *Environmental Law Reporter* 27 (August 1997): 10391; Oliver A. Houck, "TMDLs III: A New Framework for the Clean Water Act's Ambient Standards Program," *Environmental Law Reporter* 28 (August 1998): 10415.

22. See Adler et al., *The Clean Water Act 20 Years Later,* 18.

23. Ibid.

24. National Science and Technology Council, *Technology for a Sustainable Future: A Framework for Action* (Washington, D.C.: Office of Science and Technology Policy, 1994), 32.

25. See Stephen M. Meyer, "The Final Act," *New Republic* 211, no. 7 (August 15, 1994): 24–29.

26. See Daniel C. Esty and Marian R. Chertow, "A Vision for the Future," in *Thinking Ecologically,* 232.

27. See *Union Elec. Co. v. EPA,* 427 U.S. 246 (1976).

28. See Adler, "Controlling Nonpoint Source Water Pollution," 10273, explaining the state role.

29. See Thomas O. McGarity, "Missing Milestones: A Critical Look at the Clean Air Act's VOC Emissions Reduction Program in Nonattainment Areas," *Virginia Environmental Law Journal* 18 (1999): 41–101; Houck, "TMDLs,"; Houck, "TMDLs, Are We There Yet?," Houck, "TMDLs III"; David M. Driesen, "Five Lessons from the Clean Air Act Implementation," *Pace Environmental Law Review* 14 (1996): 51–62, 55–56.

30. See Kirsten H. Engel, "State Environmental Standard-Setting: Is There a 'Race' and Is It 'To the Bottom'?," *Hastings Law Journal* 48 (1997): 271–376, 340–347.

31. See Richard L. Revesz, "Rehabilitating Interstate Competition: Rethinking the 'Race-to-the-Bottom' Rationale for Federal Environmental Regulation," *New York University Law Review* 67 (1992): 1210–1254; Richard L. Revesz, "The Race to the Bottom and Federal Environmental Regulation: A Response to Critics," *Minnesota Law Review* 82 (1997): 535–564.

32. See Kirsten H. Engel and Scott R Saleska, "Facts are Stubborn Things: An Empirical Reality Check in the Theoretical Debate over the Race-to-the-Bottom in State Environmental Standard-Setting," *Cornell Journal of Law and Public Policy* 8 (1998): 55–88; Engel, "State Environmental Standard-Setting," 315–367.

33. See Engel and Saleska, "Facts Are Stubborn Things," 62.

34. See ibid.

35. See Engel, "State Environmental Standard-Setting," 275.

36. See ibid., 301–309.

37. See Brown, *State of the World 2000,* 5–6.

38. See Robert T. Watson et al., eds., *The Regional Impacts of Climate Change: An Assessment of Vulnerabilities* (Cambridge, U.K.: Cambridge University Press, 1998), 4–7, 9–15; Robert T. Watson et al., eds., *Climate Change 2001: Synthesis Report* (Cambridge, U.K.: Cambridge University Press, 2001), 60–78.

39. Brown, *State of the World 2000,* 5.

40. OECD Environmental Data: Compendium 1999.

41. Serious international discussion of the issue began in the late 1980s. See Daniel Bodansky, "The United Nations Framework Convention on Climate Change: A Commentary," *Yale Journal of International Law* 18 (1993): 451–558. The Kyoto Protocol opened for signature in 1997, but has not yet come into force.

42. See David Malakoff, "Thirty Kyotos Needed to Control Warming," *Science* 278 (December 19, 1997), 2048.

43. See Environmental Law Institute, "Ozone Depletion," *Environmental Law Reporter* 28 (1998): 10754.

44. See Stockholm Convention on Persistent Organic Pollutants, Final Act of the Conference of Plenipotentiaries of the Stockholm Convention on Persistent Organic Pollutants, Appendix II, UNEP/POPS/Conf/4 (opened for signature, May 23, 2001).

45. See Driesen, "The Societal Cost of Environmental Protection," 587–601, for a discussion of some of the theoretical and practical problems.

Chapter 9

1. See Michael R. Lozeau, "Tailoring Citizen Enforcement to an Expanding Clean Water Act: The San Francisco BayKeeper Model," *Golden Gate University Law Review* 28 (1998): 429–488, 440–441.

2. See Eileen Gauna, "Federal Environmental Citizen Provisions: Obstacles and Incentives on the Road to Environmental Justice," *Ecology Law Quarterly* 22 (1995): 1–87, 44–69.

3. See, e.g., *Gwaltney of Smithfield v. Chesapeake Bay Foundation,* 484 U.S. 49 (1987).

4. 523 U.S. 83 (1998).

5. See *Friends of the Earth v. Laidlaw Environmental Servs.,* 528 U.S. 167, 185–186 (2000).

6. See ibid., 193–194.

7. William Rodgers, *Environmental Law,* 2d ed. (St. Paul, Minn.: West Publishing Company, 1994), 99.

8. Michael D. Axline, "Decreasing Incentives to Enforce Environmental Laws: City of Burlington v. Dague," *Washington University Journal of Urban and Contemporary Law* 43 (1993): 257–274, 261.

9. See Gauna, "Federal Environmental Citizen Provisions," 76–79.

10. 505 U.S. 557 (1992).

11. See Axline, "Decreasing Incentives to Enforce Environmental Laws."

12. See ibid., 265. Justice Scalia also expressed concern about the difficulty of determining a proper lodestar fee. While this is a valid concern, surely it should not trump the need to provide adequate fees to deter violations. Furthermore, the courts or legislatures could develop firm rules of thumb to make this litigation less difficult. See, e.g., ibid., 268.

13. See Gary S. Becker, "Crime and Punishment: An Economic Approach," *Journal of Political Economy* 76 no. 2 (1968): 169–176; A. Mitchell Polinsky and Steven Shavell, "Punitive Damages: An Economic Analysis," *Harvard Law Review* 111 (1998): 869–962.

14. See Mark A. Cohen, "Empirical Research on the Deterrent Effect of Environmental Monitoring and Enforcement," *Environmental Law Reporter* 30 (2000): 10245.

15. See A. Mitchell Polinsky and Steven Shavell, "Enforcement Costs and the Optimal Magnitude and Probability of Fines," *Journal of Law and Economics* 35 (1992): 133–148.

16. See Michael S. Greve, "The Private Enforcement of Environmental Law," *Tulane Law Review* 65 (1990): 339–393, 379.

17. See ibid., 380. See, e.g., U.S. Congress, Office of Technology Assessment, "Gauging Control Technology and Regulatory Impacts in Occupational Safety and Health—an Appraisal of OSHA'S Analytical Approach," OTA-ENV-635 at 64 (1995); Nicholas A. Ashford and George R. Heaton Jr., "Regulation and Technological Innovation in the Chemical Industry," *Law and Contemporary Problems* 46, no. 3 (1983): 109–157, 109, 139–140.

18. Wendy Naynerski and Tom Tietenberg, "Private Enforcement of Federal Environmental Law," *Land Economics* 68 (1992): 28–49.

19. Some commentators have advocated using information-based strategies as a cornerstone for reforming environmental law. See, e.g., Eric W. Orts, "Reflexive Environmental Law," *Northwestern University Law Review* 89 (1995): 1227–1340. Further development of these ideas requires coming to grips with the questions I raise below.

20. See Bradley C. Karkkainen, "Information as Environmental Regulation: TRI and Performance Benchmarking, Precursor to a New Paradigm," *Georgetown Law Journal* 89 (2001): 257–370, 297.

21. Ibid.

22. See, e.g., ibid., 354–356, discussing the Massachusetts Toxic use Reduction Act.

23. See Clifford Rechtschaffen, "The Warning Game: Evaluating Warnings Under California's Proposition 65," *Ecology Law Quarterly* 65, no. 23 (1996): 303–368, 341–354.

24. See, e.g., David W. Case, "The EPA's Environmental Stewardship Initiative: Attempting to Revitalize a Floundering Regulatory Reform Agenda," *Emory Law Journal* 50 (2001): 1–100, 68 n. 417.

25. See Karkkainen, "Information as Environmental Regulation," 328.

26. See generally ibid., 327–328.

27. See, e.g., Hale E. Sheppard, "Timber Certification: An Alternative Solution to the Destruction of Chilean Forests," *Journal of Environmental Law and Litigation* 14 (1999): 301–350, 306, explaining this theory as important in efforts at timber certification.

28. Empirical evidence on this question, however, has yielded mixed results. See Karkkainen, "Information as Environmental Regulation," 326–327.

29. See ibid., 323–325.

30. See ibid., 310–311.

31. See Peter S. Menell, "Structuring a Market-Oriented Federal Eco-Information Policy," *Maryland Law Review* 54 (1995): 1435–1474, 1436, discussing the difficulty for consumers of assessing environmental choices in purchasing decisions.

32. Accord Nicholas A. Ashford, "Government and Innovation in Environmental Transformations in Europe And North America," in *Special Issue on Ecological Modernization,* ed. David Sonnenfeld and Arthur Mol, *American Behavioral Scientist* 45 (2002): 1417–1434.

33. See Klaus Lindegaar, "Environmental Law, Environmental Globalization, and Sustainable Techno-Economic Evolution," in *Environment, Technology, and Economic Growth,* ed. Andrew Tylecote and Jan van der Straaten, (Northampton, Mass.: E. Elgar, 1998), 146, claiming that firms do not maximize profits by searching in all directions for them and may inappropriately discount long term savings; Environmental Law Institute, *Innovation, Cost and Regulation: Perspectives on Business, Policy and Legal Factors Affecting the Cost of Environmental Compliance* (Washington, D.C.: Environmental Law Institute, 1999).

34. Elaine Robbins, "Good Chemistry: NRDC Teams Up with Dow to Cut Pollution and Costs," *Amicus Journal* 21, no. 3 (fall 1999): 8.

35. See James P. Karp, "Sustainable Development: Toward a New Vision," *Virginia Environmental Law Journal* 13 (1994): 239–269, 258–259; Thomas O. McGarity, "The Expanded Debate Over the Future of the Regulatory State," *University of Chicago Law Review* 63 (1996): 1463–1532, 1512 n.246.

36. See Errol E. Meidinger, "Private Environmental Regulation, Human Rights, and Community," *Buffalo Environmental Law Journal* 7 (2000): 123–238.

37. See ibid.

38. See Orts, "Reflexive Environmental Law," arguing for an approach to environmental law based on European auditing models partly because of the problem of complexity.

39. See T.H. Tietenberg, "Using Economic Incentives to Maintain Our Environment," *Challenge* 33, no. 2 (March–April 1990): 42–47, 43.

40. See Nathanael Greene and Vanessa Ward, "Getting the Sticker Price Right: Incentives for Cleaner, More Efficient Vehicles," *Pace Environmental Law Review* 12 (1994): 91–100, 94–97.

41. Ibid., 94–95.

42. See David M. Driesen, "Is Emissions Trading an Economic Incentive Program?: Replacing the Command and Control/Economic Incentive Dichotomy," *Washington and Lee Law Review* 55 (1998): 289–350, 346.

43. See Rena I. Steinzor and Linda E. Greer, "In Defense of the Superfund Liability System: Matching the Diagnosis and the Cure," *Environmental Law Reporter* 27 (June 1997): 10,286–10,295, 10,290 n. 19, citing Katherine N. Probst, Don Fullerton, Robert E. Litan, et al., *Footing the Bill for Superfund Cleanups, Who Pays and How?* (Washington, D.C.: Brookings Institution Press, 1995), 26, describing such complaints; Lloyd S. Dixon, "The Transaction Costs Generated by Superfund's Liability Approach," in *Analyzing Superfund: Economics, Science, and Law,* ed. Richard L. Revesz and Richard B. Stewart (Washington, D.C.: Resources for the Future, 1995), 171–185.

44. Klaus Lindegaard, "Environmental Law, Environmental Globalization, and Sustainable Techno-Economic Evolution," in *Environment, Technology, and Economic Growth,* ed. Andrew Tylecote and Jan van Der Straaten (Northampton, Mass.: E. Elgar, 1998), 141.

45. See ibid., 176–177.

46. See Eban S. Goodstein, *Jobs and the Environment: The Myth of a National Trade-off* (Washington, D.C.: Economic Policy Institute, 1994), 1.

47. Cf. Orts, "Reflexive Environmental Law," 1328–1329, recommending an experimental program be initiated with study of its costs after some trial.

48. See Thomas O. McGarity, *Reinventing Rationality: The Role of Regulatory Analysis in the Federal Bureaucracy* (New York: Cambridge University Press, 1991), 131.

49. See, e.g., *Pension Benefit Guaranty Corp. v. Connolly,* 475 U.S. 211 (1986), upholding requirement that private companies fund pensions for retirees after terminating a retirement plan; *Usery v. Turner Elkhorn Mining Co.,* 428 U.S. 1 (1976), upholding law requiring mining companies to compensate former employees with black lung disease; *Eastern Enterprises v. Apfel,* 524 U.S. 498 (1998), holding law applying retroactive liability to companies who never promised health protection for Black Lung Disease unconstitutional.

50. See *Eastern Enterprises,* 524 U.S. 539–47 (Kennedy concurring), 554–58 (Breyer, Stevens, Souter, and Ginsburg, dissenting).

51. See *Collins v. City of Marker Heights,* 503 U.S. 115, 125 (1992) (unanimous opinion of Stevens, J.), discussing Court's reluctance to expand the concept of substantive due process.

52. See, e.g., J. L. Lewin, "Energy and Environmental Policy Options to Promote Coalbed Methane Recovery," *Atomic Energy Commission USA—Reports* (Conf-950572) (1995), 497–508, 506, recommending subsidizing methane recovery through a methane emission fee on coal mining.

Chapter 10

1. See, e.g., Thomas O. McGarity, "Some Thoughts on 'Deossifying' the Rulemaking Process," *Duke Law Journal* 41 (1992): 1385–1462; Thomas O. McGarity, "The Courts and the Ossification of Rulemaking: A Response to Professor Seidenfeld," *Texas Law Review* 75 (1997): 525–558; Frank B. Cross, "Shattering the Fragile Case for Judicial Review of Rulemaking," *Virginia Law Review* 85 (1999): 1243–1334, 1279–1281.

2. See Cross, "Shattering the Fragile Case for Judicial Review of Rulemaking"; Jerry L. Mashaw, *Greed, Chaos, and Governance: Using Public Choice to Improve Public Law* (New Haven, Conn.: Yale University Press, 1997), 167–180.

3. See Cross, "Shattering the Fragile Case for Judicial Review of Rulemaking," 1315–1316.

4. See, e.g., Christopher F. Edley, Jr., *Administrative Law: Rethinking Judicial Control of Bureaucracy* (New Haven, Conn.: Yale University Press, 1990), 8, discussing the indeterminacy of judicial review of agency action.

5. See Cross, "Shattering the Fragile Case for Judicial Review of Rulemaking," 1270–1273, reviewing the relevant literature.

6. Cross includes other arguments not canvassed here. See ibid.

7. This may help explain social science findings that citizens sometimes fail to participate in important local regulatory proceedings. See James P. Lester and Ann O'M. Bowman, eds., *The Politics of Hazardous Waste Management* (Durham, N.C.: Duke University Press, 1983), 99, 177.

8. As an attorney with NRDC, I worked with EPA to set up these hearings, attended them myself, and worked with the local groups in Louisiana who made them successful.

9. See Lester and Bowman, *Hazardous Waste Management,* 177–189; Walter E. Rosenbaum, "The Bureaucracy and Environmental Policy," in *Environmental Politics and Policy: Theories and Evidence,* ed. James P. Lester (Durham, N.C.: Duke University Press, 1995), 228–230.

10. 424 U.S. 1 (1976).

11. 431 U.S. 209 (1977)

12. See *First National Bank of Boston v. Bellotti,* 435 U.S. 765, 794 n. 34 (1978).

13. *Buckley,* 424 U.S., 48–49.

14. See David M. Driesen, "Five Lessons from the Clean Air Act Implementation," *Pace Environmental Law Review* 14 (1996): 51–62, 53–55.

15. See Noah Sachs, "Blocked Pathways: Potential Legal Responses to Endocrine Disrupting Chemicals," *Columbia Journal of Environmental Law* 24 (1999): 289–353, 308; Stockholm Convention on Persistent Organic Pollutants (UNEP/POPS/Conf/2, Appendix II) (opened for signature May 22, 2001).

16. See David M. Driesen, "Choosing Environmental Instruments in a Transnational Context," *Ecology Law Quarterly* 27 (2000): 1–52.

17. See Charter of the International Military Tribunal, Agreement for the Prosecution and Punishment of the Major war Criminals of the European Council, August 8, 1945, 58 Stat. 1544, E.A.S. No. 472, 82 U.N.T.S. 280; Control Council Law No. 10, Punishment of Persons Guilty of War Crimes, Crimes Against the Peace and Against Humanity, December 20, 1945, 3 Official Gazette Control Council of Germany 50–55 (1946); Lawrence Douglas, "Film as Witness: Screening Nazi Concentration Camps before the Nuremberg Tribunal," *Yale Law Journal* 105 (1995): 449–481, 461–462, explaining that prosecution's theory linked the illegality of atrocities against German citizens to acts of war by the state against other states.

18. See, e.g., Convention on the Prevention and Punishment of the Crime of Genocide, 78 U.N.T.S. 277, art. IV, entered into force, Dec. 9, 1948 (requiring punishment of any person committing genocide).

19. 630 F.2d 876 (2d Cir. 1980).

20. See, e.g., *Hilao v. Estate of Ferdinand Marcos,* 25 F. 3d 1467 (9th Cir. 1994); *Trajano v. Marcos,* 978 F.2d 1439 (9th Cir. 1989); *Hilao v. Estate of Ferdinand Marcos,* 103 F.3d 767 (9th Cir. 1996); *Kadic v. Karadzic,* 70 F.3d 232 (2nd Cir. 1995); *Negewo v. Abebe-Jira,* 72 F.3d 844 (11th Cir. 1996). Cf. *Tel-Oren v. Libyan Arab Republic,* 726 F.2d 774, 776 (D.C. Cir. 1994), declining to create private cause of action against "non-state actors", but suggesting that individuals acting under color of state law would be liable.

21. U.S. Code, vol. 28, sec. 1350 (2000).

22. See generally *Torture as Tort: Comparative Perspectives on the Development of Transnational Human Rights Litigation,* ed. Craig Scott (Portland, Ore.: Hart Publishing, 2001).

23. Patricia W. Birnie and Alan E. Boyle, *International Law and the Environment* (New York: Oxford University Press, 1992), 89, characterizing this principle as "beyond serious argument".

24. Trail Smelter Case (United States v. Canada), "Arbitral Tribunal," *United Nations Reports of International Arbitral Awards,* 3 (1941).

25. See Hari M. Osofsky, "Environmental Human Rights Under the Alien Tort Statute: Redress for Indigenous Victims of Multinational Corporations," *Suffolk Transnational Law Review* 20 (1997): 335–396; Richard L. Herz, "Litigating Environmental Abuses under the Alien Tort Claims Act: A Practical Assessment," *Virginia Journal of International Law* 40 (2000): 545–638.

26. See *Aguinda v. Texaco,* 142 F. Supp. 534 (S.D.N.Y. 2001), dismissing on forum non-conveniens grounds; *In Re Aguinda,* 241 F.3d 194 (2d Cir. 2001), declining to order District Judge to recuse himself; *Aguinda v. Texaco, Inc.,* 157 F.3rd 153 (2d Cir. 1998), reversing dismissal of complaint on forum non conveniens and other grounds; *In re Union Carbide Gas Plant Disaster at Bhopal, India,* in December, 1984, 634 F. Supp. 842, 848–52 (S.D.N.Y. 1986), dismissing on forum non-conveniens grounds, aff'd as modified, 809 F.2d 195 (2d Cir. 1987); *Beanal v. Freeport-McMoRan, Inc.,* 969 F. Supp. 362 (E.D. La. 1997), dismissing case; *National Coalition Government of Union of Burma v. Unocal, Inc.,* 176 F.R.D. 329 (C.D. Cal. 1997), challenging alleged torture to facilitate a natural gas project.

27. There would remain a series of important procedural issues that would have to be resolved before such a substantive right would have practical significance. There already exists a body of scholarly literature seeking to define some acts harming the environment as human rights related crimes. See, e.g., Lynn Berat, "Defending the Right to a Healthy Environment: Toward A Crime of Geocide in International Law," *Boston University International Law Journal* 11 (1993): 327–348; Dinah Shelton, "Human Rights, Environmental Rights, and the Right to Environment," *Stanford Journal of International Law* 28 (1991): 103–138; *Human Rights Approaches to Environmental Protection,* ed. Alan E. Boyle and Michael R. Anderson (New York: Oxford University Press, 1996). I am suggesting a different approach, using transnational or international litigation against private parties to enforce general international environmental norms.

Chapter 11

1. See Environmental Law Institute, *Barriers to Environmental Technology Innovation and Use* (Washington, D.C.: Environmental Law Institute, 1998).

2. Ibid., 30.

3. See generally Nicholas A. Ashford and George R. Heaton, "Regulation and Technological Innovation in the Chemical Industry," *Law and Contemporary Problems* 46 (1983): 109–157, 137, discussing industry perception of regulatory barriers to innovation and the lack of concrete evidence to support it.

4. For an example of research that does make this distinction, see David Popp, "Pollution Control Innovations and the Clean Air Act of 1990" Working Paper #8593, NBER, 2001.

5. See Bruce A. Ackerman and William T. Hassler, *Clean Coal: Dirty Air* (New Haven, Conn.: Yale University Press, 1981), 68.

6. See Craig N. Oren, "Prevention of Significant Deterioration: Control-Compelling Versus Site-Shifting," *Iowa Law Review* 74 (1988): 1–114, 55.

7. See Richard E. Ayres and Richard W. Parker, "The Proposed WEPCO Rule: Making the Problem Fit the Solution," *Environmental Law Reporter* 22 (1992): 10201–10210, 10202, referring to the "Methuselah-like persistence of large, old power plants."

8. See, e.g., Sylvie Faucheux, John M. Gowdy, and Isabelle Nicolai, *Sustainability and Firms* (Northampton, Mass.: E. Elgar, 1998), 16.

9. See Kirsten H. Engel, "State Environmental Standard-Setting: Is There a 'Race' and Is It 'To the Bottom'?," *Hastings Law Journal* 48 (1997): 271–394, 279, 321–354.

10. See, e.g., *Wisconsin Electric Power Co. v. Reilly,* 893 F.2d 901 (7th Cir. 1991), exempting a like kind replacement project from regulation.

11. For an excellent discussion of some of the background to the relevant regulations, see Ayres and Parker, "The Proposed WEPCO Rule."

12. See Bruce A. Ackerman and Richard B. Stewart, "Reforming Environmental Law: The Democratic Case for Market Incentives," *Columbia Journal of Environmental Law* 13 (1988): 171–199, 179.

13. See Richard B. Stewart, "Regulation, Innovation, and Administrative Law: A Conceptual Framework," *California Law Review* 69 (1981): 1256–1377, 1337.

14. See Ackerman and Stewart, "Reforming Environmental Law," 180.

15. Cf. Byron Swift, "The Acid Rain Test," *Environmental Forum* 14, no. 3 (May–June 1997): 17–25, 17, arguing that emissions cap is more important to acid rain program's success than trading.

16. See, e.g., *Natural Resources Defense Council v. EPA*, No. 90-2447, 1991 WL 157261, at *1 (4th Cir. 1991); *United States v. Allsteel*, No. 87 C 4638, 1989 WL 103405, at *1 (N.D. Ill. 1989)(unpublished disposition); *United States v. Alcan Foil Products*, 694 F. Supp. 1280, 1281 (W.D. Ky. 1988), aff'd in part, rev'd in part, 889 F. 2d 1513 (6th Cir. 1989).

17. See Byron Swift, "Command Without Control: Why Cap-and-Trade Should Replace Rate Standards for Regional Pollutants," *Environmental Law Reporter* 31 (2001): 10330–10341, reviewing the history of utility rate standards.

18. See Nicholas A. Ashford, "The Influence of Information-Based Initiatives and Negotiated Environmental Agreements on Technological Change" in *Voluntary Approaches in Environmental Policy,* ed. Carlo Carraro and François Lévêque (Boston: Kluwer Academic Publishers, 1999), 137–150, stringency is the most important factor affecting technological change.

19. See Alan S. Miller, "Environmental Regulation, Technological Innovation, and Technology-Forcing," *Natural Resources and Environment* 10, no. 2 (fall 1995): 64–69.

20. See generally Nicholas A. Ashford, "Understanding Technological Responses of Industrial Firms to Environmental Problems: Implications for Government Policy," in *Environmental Strategies for Industry,* ed. Kurt Fischer and Johan Schot (Washington, D.C.: Island Press, 1993); Ashford and Heaton, "Regulation and Technological Innovation in the Chemical Industry," 154; while technology based regulation usually promotes diffusion of existing technology rather than innovation, the few regulations that go beyond that level of stringency have prompted innovation.

21. This problem seems to characterize health and safety regulation generally, not just in the environmental area. See Jerry L. Mashaw and David L. Harfst, *The Struggle for Auto Safety* (Cambridge, Mass.: Harvard University Press, 1990), 74–75, 86.

22. See *National Lime Ass'n v. Environmental Protection Agency,* 627 F.2d 416 (D.C. Cir. 1980).

23. See, e.g., Environmental Protection Agency, Final Rule, "National Emission Standards for Hazardous Air Pollutants for Source Categories; Wool Fiberglass Manufacturing," *Federal Register* 64, no. 113 (June 14, 1999): 31695, 31703.

24. See EPA Air Programs, C.F.R., vol. 40, sec. 60.44(a),(d) (1998); Environmental Protection Agency, Final Rule, "Revision of Standards of Performance for Nitrogen Oxide Emissions From Fossil-Fuel Fired Steam Generating Units; Revisions to Reporting Requirements for Standards of Performance for New Fossil-Fuel Fired Steam Generating Units," *Federal Register* 63, no. 179 (September 16, 1998): 49442.

25. See Environmental Protection Agency, Final Rule, "Revision of Standards of Performance for Nitrogen Oxide Emissions From Fossil-Fuel Fired Steam Generating Units; Revisions to Reporting Requirements for Standards of Performance for New Fossil-Fuel Fired Steam Generating Units," *Federal Register* 63, no. 179 (September 16, 1998): 49442, 49445 (arguing that utilities using high sulfur coal could meet these standards by using selective catalytic reduction).

26. See, e.g., Jerry L. Mashaw and David L Harfst, "Regulation and Legal Culture: The Case of Motor Vehicle Safety," *Yale Journal on Regulation* 4 (1987): 257–316 (1987).

27. See Christopher Flavin and Nicholas K. Lenssen, *Power Surge: Guide to the Coming Energy Revolution* (New York: W. W. Norton, 1994), 195–216.

28. See Arnold W. Reitze, Jr., "Mobile Source Air Pollution Control," *Environmental Lawyer* 6 (2000): 309–439, 412–425.

Chapter 12

1. See Joseph D. Kearney and Thomas W. Merrill, "The Great Transformation of Regulated Industries Law," *Columbia Law Review* 98 (1998): 1323–1409, 1325.

2. See ibid., 1324; changes in regulated industry frequently characterized as "deregulation."

3. See ibid., 1384–1385.

4. See, e.g., ibid.

5. See ibid., 1329–1364, describing the changes in detail.

6. See Telecommunications Act of 1996, U.S. Code, vol. 47, sec. 254 (1994)(Supp. V).

7. See ibid., sec. 254(h)(1)(B).

8. See Eli M. Noam, "Will Universal Service and Common Carriage Survive the Telecommunications Act of 1996?," *Columbia Law Review* 97 (1997): 955–975.

9. See Alfred E. Kahn, *The Economics of Regulation: Principles and Institutions* (Cambridge, Mass.: MIT Press, 1970), 190–191, arguing that cross-subsidies are inherently inefficient.

10. See Kearney and Merrill, "The Great Transformation of Regulated Industries Law," 1347.

11. See ibid., 1347–1348.

12. See ibid., 1348.

13. See ibid., 1396; Joseph Fagan, "From Regulation to Deregulation: The Diminishing Role of the Small Consumer within the Natural Gas Industry," *Tulsa Law Journal* 29 (1994): 707–734.

14. See Richard G. Newell, Adam B. Jaffe, and Robert N. Stavins, "The Induced Innovation Hypothesis and Energy-Saving Technological Change," *Quarterly Journal of Economics* 114 (1999): 941–975.

15. See Daniel Yergin and Joseph Stanislaw, *The Commanding Heights: The Battle Between Government and the Marketplace that is Remaking the Modern World* (New York: Simon and Schuster, 1998), 347. Cf. Paul Eric Teske, *After Divestiture: The Political Economy of State Telecommunications Regulation* (Albany: State University of New York Press, 1990), 6–7, collecting varying explanations for breakup of the Bell System.

16. See Kearney and Merrill, "The Great Transformation of Regulated Industries Law," 1339.

17. See ibid., 1349–1358, describing requirements.

18. See Milton M. Friedman, *Capitalism and Freedom* (Chicago: University of Chicago Press, 1962), 119–160.

19. See Kearney and Merrill, "The Great Transformation of Regulated Industries Law," 1402.

20. See William Baumol, Elizabeth E. Bailey, Robert D. Willig, et al., "Weak Invisible Hand Theorems on the Sustainability of Multiproduct Natural Monopoly," *American Economic Review* 67 (1977): 350–365; William J. Baumol, "Contestable Markets: An Uprising in the Theory of Industry Structure," *American Economic Review* 17, no. 1 (1982): 1–15; William J. Baumol, John C. Panzar, and Robert D. Willig, *Contestable Markets and the Theory of Industry Structure* (New York: Harcourt Brace Jovanovich, 1982); Elizabeth E. Bailey and William J. Baumol, "Deregulation and the Theory of Contestable Markets," *Yale Journal on Regulation* 1 (1984): 111–137.

21. See Michel Kerf and Damien Geradin, "Controlling Market Power in Telecommunications: Antitrust vs. Sector-Specific Regulation: An Assessment of the United States, New Zealand and Australian Experiences," *Berkeley Technology Law Journal* 14 (1999): 919–1020, 924–925.

22. See David M. Driesen, "Is Emissions Trading An Economic Incentive Program?: Replacing the Command and Control/Economic Incentive Dichotomy," *Washington and Lee Law Review* 55 (1998): 289–350, 303–304, 333–334; Daniel H. Cole and Peter Z. Grossman, "When Is Command and Control Efficient? Institutions, Technology, and the Comparative Efficiency of Alternative Regulatory Regimes for Environmental Protection," *Wisconsin Law Review* (1999): 887–938, 903.

23. See Kenneth R. Richards, "Framing Environmental Policy Instrument Choice," *Duke Environmental Law and Policy Forum* 10 (2000): 221–284, developing a "constrained optimization formula" that does take implementation cost into account.

Index

Acid rain
  allowance trading program, 55, 60,
    66, 67, 85, 194, 195, 197, 212
  nitrogen oxides as a contributor to,
    190
Ackerman, Bruce, 50, 51, 191, 192
Adaptive efficiency. *See* Efficiency,
    adaptive
Administrative agencies. *See also* EPA;
    Occupational Safety and Health
    Administration (OSHA);
    Rulemaking
  bounded rationality and, 7, 169
  bypassing, 176–178
  compared with private firms, 122,
    169, 175, 176
  cost-benefit analysis and, 16, 22, 23,
    27–31, 119, 210
  deference to, 42
  delegation of authority to set limits
    for polluters, 15, 16, 117, 144, 174
  emissions trading and, 61, 62
  environmental competition statute
    and, 159
  environmental metrics and, 193, 194
  judicial review and, 105, 116,
    165–167, 199
  ossification, 118, 165
  political pressures upon, 105, 116,
    119, 122, 164, 175
  pollution taxes and, 69
  public participation and, 167–171,
    176

regulatory negotiation and, 120
role in stimulating innovation,
    113–116, 165, 183, 197, 200
state agencies, 188
technology-based standards and, 50,
    51
traditional regulation and, 53
Administrative capacity, 10, 119, 124,
    125, 129
Administrative proceedings. *See*
    Rulemaking
Advertising, 9, 78
Air pollution, 77, 85, 87, 98, 124–128.
    See also Acid rain; Clean Air Act;
    Climate change; Ozone; Smog;
    Toxic substances
Alien Torts Claims Act, 180
Allowance trading. *See* Emissions trad-
    ing
Amazon.com, 6, 77, 93, 94, 102, 108
Arbitrary and capricious standard, 27,
    30, 116, 167, 169
Asbestos, 28–30, 118
Attorney fees, 142, 143, 145
Automobile. *See* Transportation

"Barriers" to innovation, 183–187
Baumol, William, 210
Beef/hormone case, 40, 41
Benzene, 29, 118
Bezos, Jeff, 6, 93, 108
Biodiversity, 39, 90. *See also*
    Endangered species

Bottlenecks, 209, 210
Bounded rationality
    of administrative agencies, 169, 210
    decisions to adopt innovations and,
        107
    defined, 7
    deregulation and, 208
    economic incentives and, 8, 57
    enforcement and, 143
    evolutionary model of innovation
        and, 95
    information strategies and, 146–148
    institutional change and, 7
    of private individuals, 47
    regulation and, 183, 185
Brazil, 133
Breyer, Stephen, 20
Burden of proof, 29, 30, 42
Bush, George H. W., 120, 146
Bush, George W., 120, 191, 208

California, 152
    high energy prices, 191, 207
California Air Resources Board, 199,
    200
Capacity. *See* Administrative capacity
Capture (of administrative agencies),
    119, 209
Carbon dioxide, 62–63, 85, 131
Carbon sequestration. *See* Forests
Carcinogens. *See* Toxic substances
Carrying capacity, 5. *See also* Optimal
    scale
Cars. *See* Transportation
Carson, Rachel, 28
CERCLA, 156, 157
Chemicals. *See* Benzene;
    Chlorofluorocarbons; Ozone; Toxic
    substances
China, 124, 133
Chloralkali plants, 53
Chlorofluorocarbons, 34, 195
Citizen suits. *See* Enforcement
Clean Air Act. *See also* Acid rain; New
    source performance standards; New
    source review

Congressional reliance upon innova-
    tion and, 113
emissions trading program for acid
    rain, 55, 176
federalism and diffuse sources, 129
new source review, 50, 187–190, 198
Clean Water Act. *See* Federal Water
    Pollution Control Act
Climate change
    affecting future generations, 71
    causes, 85, 131, 132, 190
    emissions trading under, 55, 62, 63,
        65
    framework convention, 39, 120
    global effects, 125, 131, 132
    Kyoto Protocol, 39, 120, 132
    as a prominent issue, 178
    threats to global climate, 46, 133
Clinton, William, 16
Coal
    cleaner, 66
    competition from other fuels and, 24,
        98, 99, 113, 157
    credits for construction of coal burn-
        ing power plants, 63
    environmental impacts associated
        with, 85, 192
    fuel switching from, 7, 85, 86, 153,
        201
    mining, 85, 86, 201
    washing, 51
Command and control/economic
    incentive dichotomy, 49, 56–62, 215
Command and control regulation, 49,
    50–54, 56, 57, 58, 193, 215. *See
    also* Traditional regulation
Competition and competitors. *See also*
    Environmental competition statute;
    International competitiveness; "Race
    to the bottom"
    collusion to prevent, 154, 192
    competitors as enforcers of environ-
        mental law, 153, 154, 156, 157
    contrast between competitive busi-
        ness and regulatory agency, 169

cost of regulation as stimulator of
competitor's innovation, 24
defining competitors, 155, 156, 158
diffusion of technology and, 109
economic dynamics of, 209
emissions trading and, 68, 192
equalization act and, 174
to improve the environment, 151,
152, 156, 160, 161
international decision-making and,
120, 121
internet and, 103
investment decisions and, 108, 113
perfect competition's desirability, 4
plant shutdowns and, 157, 158
pollution taxes and, 151
to provide material amenities, 152
regulated industries and, 204, 206,
207, 209, 210
regulatory decisions and, 4, 23, 24,
113, 115
regulatory design and, 192, 194
stimulating innovation, 5, 93, 94,
109, 110, 152–154, 177
in telecommunications, 209
Congress
ability to establish long compliance
time, 60, 176, 177
academic disagreement with, 130
case work and, 114, 115
citizen suits and, 140
consideration of CBA, 22
environmental goals and, 125
information strategies and, 146
104th, 16, 17
procedure for agencies and, 117
reasons for new source review,
187
regulatory criteria, 174, 199
setting specific standards for industry,
61, 62, 113, 128, 194
Congressional and agency use compared, 22
contributing to disincentives for generation of information, 155
criticisms of, 20–31, 119, 163, 212

to determine goals of environmental
regulation, 1, 3, 15, 49
economists' skepticism at time of
104th Congress, 3
emissions trading and, 55, 61
firms' use of, 94, 96
optimal deterrence and, 144
use in environmental law, 15–17
Consumption
increase, 45, 46, 102, 123–126, 129
limiting, 75
per capita, 89
Contestable market theory, 210
Convention on International Trade in
Endangered Species (CITES), 34
Corporations. *See* Firms
Cost. *See also* Cost-benefit analysis;
Cost effectiveness; Transaction costs
asbestos damages and, 28
of assessing environmental benefits,
21
automobiles and, 69
of avoiding pollution, 100
benefits from costs competitors incur,
24
change over time, 22, 200
of coal-fired power plants, 86, 98
competitiveness and, 26
computers and, 79
consumer avoidance through energy
efficiency, 25
consumption reduction and, 43
corporate avoidance of, through innovation and pollution prevention, 44
corporate profits and, 24
economies of scale and, 24
emissions trading and, 55, 56, 64, 65,
68, 193
of enforcement, 144, 212
environmental competition statute
and, 153, 154, 156–159, 213
equalization act and, 172, 175
evenness of, 19, 54
of existing technology, 65
failure to internalize, 8, 10
falling costs over time, 86, 104

Cost (cont.)
  fears of high costs impeding environmental goals, 86
  free market imagery and, 18
  free trade and, 37, 43–45
  of fuel switching, 153
  inability to compare with environmental effects, 21
  innovation and, 75, 78–87, 91, 94, 97, 104, 109–112, 133, 158, 184, 197, 200, 208, 213
  of insulation, 101
  long-term, 5, 65, 84–87
  magnitude of pollution control costs, 26
  marginal, 5, 44, 69–71, 188
  need to consider public sector costs, 212
  need to decrease emphasis upon, 12, 213
  new source review and, 187–189
  opportunity costs, 100, 212
  optimal regulation and, 23
  overestimation of, 22, 23, 31, 69, 159, 164, 213
  performance standards and, 52, 67, 104
  plant shutdowns and, 16, 120
  polluter pays principle and, 19
  pollution taxes and, 54, 68, 69
  prevention cost, 15, 18, 21
  of raw materials, 99, 100
  reduction in response to regulation, 26
  of road building, 79
  role in economic dynamic reform generally, 212, 213
  savings from deregulation, 206
  savings from pollution prevention, 145, 149
  short-term, 67, 83, 84, 87, 101
  siting decisions and, 188
  social, 18, 23, 24, 68, 69
  of solar power, 98
  statutory requirements to consider, 16, 114, 200
  treaties and, 34, 129
  uncertainty about, 109, 133, 159
  value of, 23–27, 31
  of water treatment, 128
  of zero emissions vehicles, 65
Cost-benefit analysis
  allocative efficiency rationale for, 4, 11, 13, 15, 17–22, 71, 211
Cost effectiveness, 17, 49, 54, 55, 71, 86
Cotton dust, 53
Courts. *See* Judicial review
Creativity, 4, 5, 73, 78
Cross, Frank, 165–167
Customary international law, 180

Daly, Herman, 44–47, 89, 196
Decentralization
  of environmental problems, 134
  of free markets, 95, 108, 110, 119
  of government and markets compared, 10, 122, 135, 181
  of standard setting, 140, 145–152
Decision-making structures
  free market, 11, 107–112, 115, 116, 119, 161, 163, 165, 175, 181, 215
  government, 11, 12, 112–122, 161, 163, 165, 175, 181, 201, 215
Deforestation. *See* Forests
Demand
  for electricity, 25
  for environmental improvement, 177
  for environmental innovation, 104, 112, 117, 118, 152, 200
Dematerialization, 102, 103
Deregulation. *See* Privatization
Design of regulation. *See* Regulatory design
Diffusion (of technology), 76, 103, 125, 133
Dow Chemical, 94, 149
Drive plus program, 152
Dynamic efficiency. *See* Efficiency, dynamic

Eastern Europe, 100
Ecological economics, 11

Economic development, 5, 45–47, 196, 197

Economic dynamics
  adaptive efficiency and, 159
  contrast with efficiency-based approach, 5, 11–13, 70–71, 75, 137, 158, 163, 213
  cost-benefit analysis and, 22–31
  definition, 6–9, 205
  distinguished from dynamic efficiency, 11, 71
  economic incentive programs and, 56–70, 212
  focus on broad systems, 27, 125
  focus on change over time, 13, 27, 45, 47, 75, 178, 208
  of free markets, 4, 9, 11, 23, 44, 137, 151, 165, 177, 181
  free trade and, 43–45, 47
  of information, 147–151
  of innovation, 4, 6, 11, 12, 45, 49, 56, 58, 68, 70–81, 90, 93, 116, 134–137, 151, 161, 208
  of international environmental law, 178, 179, 181
  of law and government decision-making, 6–9, 11, 22, 23, 27, 30, 31, 70, 114–116, 125, 132, 135, 137, 139, 140, 145, 163, 165, 169–171, 175–178, 181, 196, 201
  of new source review, 191
  pace of environmental decision-making and, 196
  of privatization, 140, 143–151
  public choice theory and, 8, 11, 114–116, 135, 140, 163, 165, 169–172
  questions raised by analysis of, 4, 6, 10–13, 23, 43, 47, 57, 75, 93, 135–138, 161, 163, 176
  regulated industries and, 207–209
  of regulatory "barriers," 188, 197
  regulatory design and, 183, 193, 194, 200
  role in legal theory, 203, 205, 210, 215, 216

  shaping consumer preferences and behavior, 90
  of traditional regulation, 183–187, 197, 200
  value of economic dynamic analysis, 5, 8, 10–12, 31, 75, 91, 151, 163, 215

Economic growth
  causes of, 4, 5, 75, 83
  in developing countries, 133
  emissions trading and, 60
  limits to, 44–47
  regulation and, 16, 60, 196, 215

Economic incentives. *See also* Moral incentives
  continuous, 9, 11, 53, 56, 58–61, 63, 70, 93, 95, 98, 103, 105, 151, 154, 181
  to enforce regulation, 140–145
  generally, 8
  to harm the environment, 97, 98, 102, 103, 126
  to improve environment, 99, 100, 103–105
  information and, 146–151
  for innovation, 93–105, 107, 139, 152, 153, 215
  negative, 58, 93, 152
  pollution taxes and, 54, 68
  positive, 58, 68, 93, 152
  power of, 2, 52
  programs, 1, 11, 13, 49–71, 183, 215
  to use cheap innovations to comply with performance standards, 52

Economic regulation. *See* Regulated industries

Economics. *See* Ecological economics; Institutional economics; Law & Economics; Neoclassical economics

Economies of scale, 24

Ecosystem services, 98

Effects-based regulation. *See* Health-based regulation

Efficiency. *See also* Cost effectiveness; Energy efficiency
  adaptive, 7, 96, 97, 130, 131, 133, 134, 148, 159, 166

Efficiency (cont.)
  allocative, 3, 5, 13, 17, 19, 20, 27,
    36, 44, 49, 71, 103
  competing efficiencies, 210–212
  dynamic, 11, 71
  economic, 3, 4, 86, 150, 158, 163
  free trade and, 33, 43, 44, 47, 48
  generally, 11, 13, 73, 137
  ideal, 1
  inefficiency, 5, 54, 55, 101, 118
  intertemporal, 4
  Kaldor-Hicks, 18
  short term, 65, 67
  static, 3–5, 11, 20–22, 75, 205–213
  tension with freedom creativity and
    growth, 96, 97
  too much emphasis upon, 213
  view of innovation and, 79, 95
Electric cars. *See* Zero emission vehicles
    (ZEVs)
Electricity
  consumption and consumer cost, 24,
    25
  costs as inducement to conserve, 101,
    149
  deregulation and increased cost in
    California, 207
  generation as a pollution source 24, 103
  innovations in generation to reduce
    pollution, 76, 86, 98
  production cost, 99
  provision of as a natural monopoly,
    208
Electric utilities
  climate change and 63, 120
  coal-fired, 7, 24, 63, 85, 190
  deregulation, 2, 98, 206
  economic dynamic influencing regula-
    tion of, 114
  emissions trading and, 63, 65, 67,
    192, 194, 195
  energy efficiency and inefficiency, 100
  fuel choice, 98, 99, 199
  new source review, 190, 191
  nitrogen oxide emissions, 127, 131,
    190

  nuclear power plants, 157
  output, 103
  regulation of, 87, 156, 157, 164, 191,
    192, 194
  siting of, 191
  sulfur dioxide emissions, 176
Emissions trading
  dependence on government decision-
    making, 59–63, 70, 152, 154, 159,
    176
  design of, 63, 191–196, 200
  efficiency of, 54–56, 67, 68, 70, 211,
    212
  innovation and, 11, 57–70, 104, 185,
    191–196
  as means of meeting environmental
    goals, 1
Employment. *See* Unemployment
Endangered species
  free market's role in causing the prob-
    lem, 100, 101
  international attention to, 178
  resistence to technological solutions, 89
  trade restrictions and, 34, 35
  World Trade Organization and, 46
Endangered Species Act, 129
End-of-the-pipe controls. *See also*
    Command and control regulation;
    Performance standards; Technology-
    based regulation; Traditional regula-
    tion
  costs, 83, 84
  pollution prevention and, 186, 194,
    197, 199
  shift away from, 36
Energy
  efficiency, 25, 98, 100, 101, 112, 150,
    152
  conservation, 101, 149, 150
  consumption, 103, 133
  services, 24
  sources, 24, 65, 67, 85
Enforcement
  citizen suits, 140–145, 154
  environmental competition statute
    and, 154

Engel, Kirstin, 130
Entrepreneurs
  environmental, 12, 110, 111, 113,
    115, 119, 120, 163–165, 170, 175,
    213
  incentives for material and environ-
    mental entrepreneurs compared, 9,
    104, 107
  material, 94, 95, 108–110, 138
Environmental competition statute,
    151–161, 163, 213
Environmental Defense, 149
Environmental groups. *See also*
    Environmental Defense; Natural
    Resources Defense Council; New
    Jersey Public Interest Research
    Group
  Forest Stewardship Council, partici-
    pants, 151
  litigation, 116, 117, 165, 192
  new source review, objections to
    relaxation, 191
  participation in regulatory proceed-
    ings, 115–117, 120, 164, 171–175,
    213
  radical substitutes, advocates for, 121
Environmental justice, 21, 22, 143,
    170–172
Environmental Law Institute study (of
    innovation), 184
Environmental metrics. *See* Metrics,
    environmental
EPA. *See also* Administrative agencies;
    Rulemaking
  air quality standards, 127
  allocation (pre-cleanup) of responsi-
    bility for pollution under Superfund,
    157
  authority to regulate, 15, 30
  compliance methods, statutory
    restrictions on authority to mandate,
    50
  cost-benefit analysis and, 16, 28–31
  emissions trading programs, 55, 194
  enforcement, 141
  health-based criteria, use of, 15

information-based programs and,
    146, 150
large-scale source regulation,
    126–128
"mass-based" versus "rate-based"
    limitations on pollution, 195
mercury, regulation leading to process
    innovations, 53
new source review, 192, 193,
    197–199
opportunistic regulation, 164
pollution tax, 69
public participation and, 167–171
reducing need for administrative deci-
    sion-making, 177
traditional regulation using standard
    industrial classification (SIC) codes,
    155
vehicle emissions, regulation, 128
vinyl chloride, regulation of, 53
water treatment, cost estimate, 128
Equalization Act, 171–175
Equilibrium, 4, 5, 23, 59, 208
Equity. *See* Environmental justice
European Union, 35, 37, 40, 127, 131
Experimentation
  adaptive efficiency and, 7, 96, 97,
    130, 159, 166
  with emissions trading, 55
  free market encouragement of, 137,
    166
  as generator of innovation and
    wealth, 4, 5, 94, 96
  upon human beings, 42
  regulation and, 51, 131, 135, 166
Externality, 18, 98, 99, 153

Federal Insecticide Fungicide and
    Rodenticide Act (FIFRA), 16, 17,
    27, 28, 30
Federal Water Pollution Control Act,
    114, 128, 129, 141
FedEx Corporation, 94
Firms, 7, 109, 115, 164, 175, 213
Fisheries, 35, 101, 152
Fixed level standards, 194, 195

Forests
  carbon sequestration, 131, 132
  damage from harvesting, 100, 124
  eastern, 130
  growth, 63
  old growth, 130
  rainforest, 90, 131, 132
Forest Stewardship Council, 150, 151
Formaldehyde standard, 53
Fossil fuels, 88, 131
Free market. *See also* Economic incentives; Market emulation
  competition and, 151–154
  creation of wealth and power and, 8
  decentralized nature, 10, 110, 119,
    122, 137, 181
  decision-making in, 11, 110–112, 119,
    122, 151, 166, 177, 181, 188, 191
  economic growth and, 5
  efficiency of, 3, 18, 21, 137, 152,
    153, 158
  environmental destruction and, 8, 9,
    97, 98, 105
  hypothetical, 18, 137
  ideology, 185
  incentives for innovation, 5, 6, 9, 73,
    77, 94–99, 103–105, 110–113, 119,
    130, 137, 151–154, 158, 166, 181,
    184, 185
  limited incentives for environmental
    improvement, 98–101, 137, 147,
    153, 184
  perfect market model, 3, 137
  regulatory stimulation of market in
    pollution control, 57, 103–105
Free speech, 172–174
Free trade, 1, 11, 13, 33–47, 89
Friedman, Milton, 209
Fuel cells
  competition with coal, 24, 113, 164
  improved environmental quality, 85,
    86
  as an innovative technology, 65, 76,
    86
  LEV program's encouragement of,
    200

Future, need to account for. *See* Time,
  change over
Future generations, 21, 71

Gasoline. *See* Oil
GATT, 37–39
General purpose technologies, 81–83,
  85, 86
Goals (of environmental programs)
  consumer goals served by polluting
    industries, 25
  cost-benefit analysis and, 1, 15, 17,
    20, 27, 71, 125, 154
  equitable, 16
  failure to achieve, 129, 130, 144, 145
  how available means affect selection
    of ends, 86
  innovation and, 75, 77, 83, 85, 86,
    91
  means of achieving, 10, 16, 49, 57,
    58, 65, 68, 69, 71, 141, 196, 197
  protecting public health and the envi-
    ronment, 15, 16, 125
  sustainable development, 45, 89
  symbolic, 114
Government agencies. *See*
  Administrative agencies
Green Lights program, 149
Greve, Michael S., 144

Hard look review, 61, 116, 166, 167
Hazardous substances. *See* Toxic sub-
  stances
Health-based regulation, 15–17, 61,
  113
Heinzerling, Lisa, 19
Human rights, 180
Hydroelectric power, 24

India, 133
Induced innovation hypothesis, 208
Inefficiency. *See* Efficiency, inefficiency
Informal rulemaking. *See* Rulemaking
Information
  lack of, 7, 155
  lack of perfect information, 150

as a regulatory tool, 145–150
Innovation
contribution to economic growth,
    4–6, 26
definition of, 75–77
degrading the environment, 9, 123,
    215
direction of, 89, 93
economic dynamics and, 75
efficiency, 65, 67 73, 161
emissions trading and, 56–68
environmental, 11, 73, 78, 80–87, 91,
    93, 99, 104–107, 110–133,
    117–122, 134, 135, 164, 184–187,
    196, 197, 200, 215
financing, 104, 107–111, 122
free markets and, 110
incremental, 80, 81, 113, 117
influence upon economic regulation's
    future, 208
limit to its importance, 89–91
material, 11, 77, 78, 81, 82, 93, 97,
    98, 103–107, 110, 111, 117–119,
    132, 135, 138, 176, 200
new source review and, 187, 188
paths of, 77
performance standards and, 50–52
pollution taxes and, 69, 70
private certification and, 151
qualitative,78–80, 85, 91, 96, 111,
    135
quantitative, 78–80, 85–87, 91,
    135
radical, 94, 96, 73, 80–87, 91, 105,
    113, 117, 133, 164
stringency and, 197
sustainable development and, 89
traditional regulation and, 49–54,
    103, 104, 183–187
typology, 77–82
value of, 11, 75–91, 133, 135
Institutional economics, 7, 11
Integrated pest management, 155
International competitiveness, 25–27
International environment court,
    178–180

International environmental law. *See*
    Climate change, framework conven-
    tion, Climate change, Kyoto
    Protocol; Convention on
    International Trade in Endangered
    Species (CITES); Customary interna-
    tional law; Montreal Protocol on
    Substances That Deplete the Ozone
    Layer; Treaties
International trade. *See* Free trade
Internet, 6, 77, 93, 94, 102–104, 155
Intertemporal spillover, 87
Irreversibility, 132
ISO standards, 150, 151

Jobs. *See* Unemployment
Joint implementation. *See* Climate
    change, emissions trading under
Judicial review, 27–31, 42, 105, 116,
    117, 165–168

Kearney, Joseph D., 207
Kyoto Protocol. *See* Climate change,
    Kyoto Protocol

Law & Economics
core ideas of, 2–4
economic dynamics theory's critique
    of, 5, 6, 11, 203, 216
influence of, 1, 2, 144, 206
Lead phase down rule, 66, 126, 197
Legal theory, 205
Lime kiln standards, 198
Lobbying, 9, 25, 114, 160, 185, 200
Lock-in of technologies, 88, 89
Louisiana, 124, 170
Low emission vehicles (LEVs), 64, 65,
    111, 199, 200

Macroeconomics, 4, 10, 43
Marine Mammal Protection Act, 38
Market emulation
analysis of market virtues, 71
efficiency-based reforms, 152
as a policy goal, 2, 3, 5, 9, 56, 176
Mashaw, Jerry, 165, 166

Mass-based limits, 60, 67, 195–197
McDonald's, 149
McGarity, Thomas, 118, 165
Merrill, Thomas W., 207
Metrics, environmental, 193–200
Methane, 131
Methyl bromide, 126
Microeconomics, 1
Montreal Protocol on Substances That
    Deplete the Ozone Layer, 34, 37,
    121, 126, 180
Moral incentives, 147, 148

National autonomy principle, 35, 36, 180
Natural gas, 7, 24, 67, 164, 199, 209
Natural monopolies, 208–210
Natural resources
  depletion of, 5, 97, 101
  renewable and nonrenewable, 9, 45
  reducing use of, 89, 100
Natural Resources Defense Council
    (NRDC), 141, 149, 170
Negotiated rulemaking, 119, 120, 168
Neoclassical economics, 1, 11, 19, 44,
    86
New Hampshire's proposed industry
    average performance system, 152
New Jersey, 170
New Jersey Public Interest Research
    Group, 141
New source performance standards,
    50, 51, 198, 199
New source review, 187–192
New York (state), 190
New Zealand, 152
Nitrogen oxides
  from coal-fired power plants, 85
  emissions fees and trading, 156, 157,
    192
  environmental effects, 127, 131, 190
North, Douglass, 7
Notice and Comment. *See* Rulemaking
Nuclear power, 82, 157

Occupational Safety and Health Act,
    29

Occupational Safety and Health
    Administration (OSHA), 15, 29, 53
Office of Management and Budget
    (OMB), 16, 19
Oil
  drilling, 86
  refinery emissions, 85, 87, 120, 127
  spills, 85–87
  vehicle pollution, 85, 88
  water pollution and, 128
Optimal enforcement, 143–145
Optimal pollution, 17, 19, 27–31, 125,
    130. *See also* Cost-benefit analysis;
    Efficiency, allocative; Optimal regu-
    lation
Optimal regulation, 23, 27–31, 213.
    See also Optimal pollution
Optimal scale, 44–46
Optimization, 95
Organic agriculture, 111, 148
Organizational theory, 7
Ossification (of rulemaking), 118
Ozone
  stratospheric, 34, 35, 53, 121, 126,
    133, 176, 178, 180, 195, 197
  tropospheric, 127, 190 (*see also*
    Smog)

Pareto optimality, 21
Particulate, 190
Path dependence, 7, 80, 88, 89, 205
Percentage reduction standards, 193,
    194
Perfect competition. *See* Competition
    and competitors, perfect competi-
    tion's desirability
Perfect information. *See* Information,
    lack of perfect information
Performance standards, 50–52
Persistent Organic Pollutants (POPS)
    Convention, 133
Persistent pollutants, 7, 132, 133, 178
Pesticides
  alternatives to, 148, 155, 156
  environmental effects, 28, 34, 128
  regulation of, 17, 28, 111, 126

Pigovian tax, 18. *See also* Tax on pollution
Polluter pays principle, 15, 17, 31
Pollution prevention
  costs and, 83, 84, 100, 149, 186
  information strategies and, 146,
    148–150
  motivation for, 147
  regulation and, 186, 187, 194–199
  significance for international interdependence, 36
Pollution from ships, 35
Pollution sources. *See* Sources of pollution
Popp, David, 67
Population growth
  environmental improvements, in spite
    of, 16
  environmental law and, 129, 215
  increased consumption, cause of, 9,
    25, 45, 46, 102, 103, 123–125
  limiting, 75, 89
  sustainable development and, 46
Porter hypothesis. *See* Porter, Michael
Porter, Michael, 26, 44
Posner, Richard, 3
Power plants. *See* Electric utilities
Precautionary principle, 30, 40
Preferences, private, 20–22, 90
Prevention cost, 18
Priority setting, 3
Privatization
  common carrier deregulation, 206
  economic dynamic analysis, 201, 215
  environmental law and, 12, 137,
    139–163, 183
  increasing use of, 2
  utility deregulation, 192, 206, 207
Proposition 65, 146
Public choice, 11, 55, 56, 114
Public good. *See* Externality
Public interest, 1, 2
Public participation, 117, 167–172
Public/private distinction, 113, 139
Public utilities. *See* Electric utilities

Quantification (of health and environmental effects), 21, 29, 30, 41, 133

"Race to the bottom," 129, 130
Radioactive waste, 87
Rainforests. *See* Forests, rainforest
Rate-based standards, 67, 195, 196
Reagan, Ronald, 16, 121, 141
Regulated industries, 203–210
Regulatory design, 12, 138, 183–201,
  215
Regulatory negotiation. *See*
  Negotiated rulemaking
Regulatory reform
  analytical requirements, addition of,
    118, 119
  cost-benefit analysis as a component
    of, 20, 31
  defense of status quo as a response to,
    1
  economic dynamics, as analytical
    tool, 6, 10, 12, 181
  economics-based, 1
  efficiency, as a goal of, 54, 75, 144,
    183, 205, 211, 212
  industry and, 2, 20
  information needed for, 146, 176
  legislation, 16, 17
Renewable energy, 67. *See also* Fuel
  cells; Hydroelectric power; Solar
  energy
Renewable resources. *See* Natural
  resources
Reproductive toxicity. *See* Toxic substances
Reputational incentives, 147
Responsible care program, 150
Revesz, Richard, 129, 130
Ricardo, David, 36
Right to pollute, 18
Risk assessment, 28–31, 40–42
Rodgers, William H., 142
Rulemaking, 22, 115–121, 140,
  167–172, 176, 207
  bypassing, 176–178

Satisficing, 95
Scale. *See* Economies of scale; Optimal
    scale
Shrimp/turtle case, 38, 46
Shutdowns (of facilities) 16, 62, 113.
    See also Unemployment
Silent Spring, 28
Simon, Herbert, 95
Smith, Adam, 36
Smog, 27, 86, 127, 190
Social cost. *See* Cost, social
Soft variables, 21
Solar energy, 65, 76, 98, 99, 164
Sources of pollution. *See also*
    Transportation
  large, 126
  multiple, 27, 123, 126, 128–131, 168
Sovereignty. *See* National autonomy
    principle
Soviet Union (former), 100
Spatial flexibility, 64–66
Spillover from innovation, 95. *See also*
    Intertemporal spillover
SPS agreement, 40–42
Standing, 142
Steady-state economies, 46, 47
Stewart, Richard B., 52, 77, 191, 192
Stringency (of environmental regula-
    tions)
  barriers to, 61, 197–201
  criteria determining, 15, 16
  innovation, effect on, 103, 104, 112,
    197–201
  potential supporters of stringent regu-
    lation, 115, 117
Substantial evidence test, 29, 30
Substantive due process, 160
Substitute goods and substances, 45,
    90, 101, 197
Sulfur dioxide, 50, 53, 55, 85, 176
Superfund law. *See* CERCLA
Sustainable development
  definition, 45–47
  economic dynamic analysis, 80
  environmental regulation and, 138,
    196, 215

  population reduction, role of, 89
  technological innovation, role of, 88,
    89
  World Trade Organization endorse-
    ment of, 46
SUVs, 82
Swift, Bryon, 67

Takings, 160
Tax on pollution, 54–56, 68–70, 151,
    152, 154
Technological innovation. *See*
    Innovation
Technology-based regulation. *See also*
    Command and control regulation;
    Traditional regulation
  available technology, industry use of,
    113, 114
  command and control, characteriza-
    tion as, 50, 51
  complexity claim about, 53
  cost-benefit criterion, as substitute
    for, 17
  environmental improvements result-
    ing from, 16
  pollution tax, similar information
    required for decision-making, 69
  trading programs, as criterion for, 61
Technology forcing regulation, 105,
    114, 165
Telecommunications Act of 1996, 206,
    207
Temporal dimension of environmental
    law and problems. *See* Time, change
    over
Thailand, 124
Thermodynamies, 9
Time, change over
  economic dynamic policy implica-
    tions, 5–7, 11, 43, 47, 178, 208,
    213, 215
  efficiency-based reforms, 13
  optimal scale theory, 45
  rates of, 134
  technology deployment, 76
Tort, 28

Toxic release inventory, 146, 148
Toxic substances. *See also* Asbestos;
    Benzene; Formaldehyde standard;
    Toxic Substances Control Act
    (TSCA)
  bioaccumulation, 28
  carcinogenic growth hormones, 40, 41
  computers and, 103
  cost-benefit analysis in regulation, 28
  pollution prevention and, 83, 197
  public response to emissions, 146,
    170
  reproductive toxicity, international
    response, 178
  sources, 83, 85, 133
  Toxic Release Inventory (TRI), 146,
    148
  waste sites regulated under
    Superfund, 156, 157
Toxic Substances Control Act (TSCA),
    16, 27–30, 165, 198
Trade restrictions, 34–36, 44, 178
Traditional regulation
  criteria for setting standards, 61
  critique of, 53, 61, 183
  dependence on government decisions
    regarding emissions, 59
  emissions trading, comparison with,
    54, 56–68, 211
  environmental competition law, com-
    parison with, 154
  erroneous characterizations as barrier
    to innovation, 49–54, 183, 184
  pollution taxes, comparison with, 69,
    70, 152
  regulatory design and, 192–195, 215
  regulatory reform and 1, 183
Trail smelter arbitration, 180
Transaction, 22, 94, 134, 181, 211
Transaction costs, 18, 27, 156, 157
Transboundary pollution, 180, 181
Transnational environmental law and
    litigation, 180, 181
Transportation. *See also* Low emission
    vehicles (LEVs); Zero emission vehi-
    cles (ZEVs)

automotive innovations, 81, 82, 85,
    86, 200
  emission reductions, 85, 128
  energy inefficiency, 98
  hazards of automobile travel, 85
  Internet as a source of potential effi-
    ciencies in, 86, 103
  lobbying of automotive industry, 120
  manufacture of autos, 120
  as a material good, 9, 76, 78, 80, 82,
    102, 111
  of oil to market, 86
  ozone, damage to cars, 127
  pollution from automobile use, 78,
    82, 85, 125, 127, 128, 131
  regulation of (including associated
    pollution), 87, 117, 128, 152, 176,
    177
  related industries, creation of, 88
  social change resulting from automo-
    bile use and development, 77, 79,
    80, 81, 88, 90
  vehicle fees/rebates, in California,
    152
Treaties, 33–35, 120–122, 178–181
Tuna/dolphin case, 38, 44

Uncertainty, 7
  business, 94, 154
  citizen suits and, 142, 145
  cost-benefit analysis and, 25, 96, 109
  implementing treaties, 122
  increasing uncertainty to spur pollu-
    tion reduction, 177
  innovative, 95, 104, 130, 133, 160,
    166
  and revision of pollution limits, 61,
    104, 119
  scientific, 42
  and setting pollution tax rates, 69, 70
  and technology-based regulation, 54
Unemployment, 25, 27, 137, 158, 199
Unfunded Mandates Act, 16
Unilateral action, 35, 38, 39
Universal service, 206
Utilities. *See* Electric utilities

Values, public, 20, 21
Vehicles. *See* Transportation
Volatile organic compound, 127

Water pollution, 85, 124, 125, 128,
    129
Wealth maximization, 4, 5
Wetlands, 97, 128, 129
Windmills, 24, 76
World Trade Organization (WTO),
    38–46

Zero emission vehicles (ZEVs), 65, 76,
    197, 200